Rüdiger Hoffmann, Jürgen Trouvain (Eds.)
HSCR 2015
Proceedings of the First International Workshop on the
History of Speech Communication Research
Dresden, September 4-5, 2015

TUD*press*

Studientexte zur Sprachkommunikation
Hg. von Rüdiger Hoffmann
ISSN 0940-6832
Bd. 79

Rüdiger Hoffmann, Jürgen Trouvain (Eds.)

HSCR 2015

Proceedings of the First International Workshop on the History of Speech Communication Research

Dresden, September 4-5, 2015

TUD*press*

2015

Bibliografische Information der Deutschen Nationalbibliothek
Die Deutsche Nationalbibliothek verzeichnet diese Publikation in der
Deutschen Nationalbibliografie; detaillierte bibliografische Daten sind
im Internet über http://dnb.d-nb.de abrufbar.

Bibliographic information published by the Deutsche Nationalbibliothek
The Deutsche Nationalbibliothek lists this publication in the Deutsche
Nationalbibliografie; detailed bibliographic data are available in the
Internet at http://dnb.d-nb.de.

ISBN 978-3-95908-020-0

© 2015 TUDpress
Verlag der Wissenschaften GmbH
Bergstr. 70 | D-01069 Dresden
Tel.: +49 351 47969720 | Fax: +49 351 47960819
http://www.tudpress.de

Preface

The international workshop on the "History Speech Communication Research" is the first one of this kind organised by the Special Interest Group (SIG) on "The History of Speech Communication Sciences". This SIG is supported by the International Phonetic Association (IPA) and the International Speech Communication Association (ISCA). This workshop which will be a unique exchange forum for researchers with work on all kind of historical aspects of the research fields represented at the Interspeech conferences and the Congresses of Phonetic Sciences (ICPhS).

This first workshop is a satellite event of Interspeech 2015 which immediately takes place after the workshop in Dresden. In this context we have also the honour to support the main conference on their website with a selection of short journalistic contributions as a historical review on a monthly basis.

Integrated in the workshop is the re-opening of the historical acoustic-phonetic collection (HAPS) in new rooms. This event is preceded by the invited opening lecture by John Ohala (University of Berkeley) on "A brief history of Experimental Phonetics in the 18th and 19th centuries". With the Barkhausen building on the first day of the workshop and the technical collections of the city of Dresden in the historical Ernemann building on day two of the workshop we have also appropriate locations for our meeting.

The contributions for this workshop contain various topics on the field: from "mechanical speech synthesis" over "collections" up to "pioneering work in phonetics". The more than dozen contributions collected in these proceedings give an encouraging signal that shows that historical aspects in our research communities can represent more than a special session at a conference. We hope that the first workshop on the "History of Speech Communication Research" will have second and more follow-up events.

Rüdiger Hoffmann and Jürgen Trouvain

Dresden and Saarbrücken, July 2015

Contents

Opening session

Mechanical speech synthesis

Collections

Pioneering work in phonetics

The Collections of the TU Dresden – Places for Teaching and Research

Greeting of the Director of the Kustodie

Kirsten Vincenz

Technische Universität Dresden, Kustodie
kirsten.vincenz@tu-dresden.de

As one of the oldest institutions of technical education in Germany, the TU Dresden can look back at a long tradition of technical training and research. Since the founding of the university in 1828, models, instruments and other scientific objects have been collected and used for educational purposes. These objects were mostly bought, or self-produced, to train students, to be used as illustrative examples for lectures or for research purposes in various different fields, and they have been integral, important tools for propagating the understanding of science. Today the TU Dresden possesses about 40 collections. According to the specific demands of a technical university, most of them originated in the technical disciplines or the natural sciences, but there is also a substantial art collection with about 3500 works of art. Objects in university collections are still considered a source of knowledge, and some of them have only recently been rediscovered as a resource for teaching and research. Over time, as disciplinary practices have developed, certain objects or even entire collections have moved away from their original purpose and become primarily objects of historical and cultural value. As such, they remain important material witnesses to the development of specific scientific disciplines and are inextricably tied to the history of the university.

The Kustodie at the Technical University was established in 1979 as a department dedicated to the preservation of not just these objects but university history as a whole, and soon after its founding, it formulated the first university policy concerning the stewardship of the collections. The main task of the Kustodie is to preserve the scientific and cultural heritage of the university and to provide the necessary infrastructure so that the collections can be used in research, teaching and as cultural symbols of the university as it has grown over the course of its nearly two century-long history. As an independent administrative unit of the university, the Kustodie provides support to the institutes and faculties that hold collections, by setting forth long-term objectives for the conceptual shaping of the collections and offering assistance to curators in the management and digitalization of artifacts according to museum standards. Furthermore, the Kustodie oversees the "ALTANAGalerie, University Collections Art plus Science", an exhibition space in the galleries of the Görges-Bau, the Electrotechnical Institute. The conceptual focus of the biannual exhibitions held in this space is the link between science and artistic practices. The exhibitions also attempt to function as an interface between the ways the material collections have been used to produce knowledge in the past and the ways they may be used in the future.

The Historic Acoustic-Phonetic Collection (HAPS) at the Institute of Acoustics and Speech Communication is one of the most outstanding and best-preserved collections of the Technical University. The fate of most university collections is determined by the interest of the caretaking institutions and chairs, for they are put to best use within their original academic field and under the care of dedicated professionals. For many years now, the HAPS has been under the guardianship of Professor Rüdiger Hoffmann, which is a very fortunate circumstance for our university and all professionals interested in the history of phonetics and speech synthesis. As a true expert in the field, he was able, together with Professor Dieter Mehnert, to build a collection unrivaled in Europe, which covers the material basis of the field of speech synthesis, speech recognition and experimental phonetics. It is also thanks to Professor Hoffmann that we can celebrate the reopening of the collection at a new location. The HAPS is of great interest not only to scholars but also to the general public; the collection is engaged in a variety of public activities, such as conferences, exhibitions, guided tours and open campus days. In addition to this volume, there have been already three publications dealing with the collection and related historical research topics. Moreover, there will be a lengthy, illustrated article by Professor Hoffmann in our second anthology about the scientific and art collections of the TU Dresden, which will also be published in September of this year. The new rooms will support a variety of activities as well as provide the space for special research and structured collection management. Although he retired last year, Professor Hoffmann has agreed to remain the caretaker of the collection, and we are very grateful for his past and future commitment to the HAPS collection and to the TU Dresden. And as it seems that in Professor Peter Birkholz, he has already found a young successor with an interest in the collection as a historic treasure, a means for teaching and a material basis for new projects.

Over the past few years, university collections have gained more attention on both the national and international level. While the situation of many collections can be improved with regard to their lack of resources and professional staff, university collections are being recognized anew for their value to the scientific community as a resource with unlimited potential for research. In 2010, the first German conference on university collections took place in Berlin and has since become an important annual event for collection professionals. The National Coordination Centre for University Collections, funded by the Federal Ministry of Education and Research, was set up only two years later. And finally, the Society of University Collections was established the same year and now functions as the co-organizer of the annual conference. This process of self-organization has given important impetus for the universities to recognize the value of their collections but also has had an impact on a political level. Since 2010, several governmental funding lines have been established to support university collections and their on-going contributions to the production of knowledge. This year, the technical universities in Freiberg and Dresden have united to co-host the national conference for university collections in order to showcase the specific nature and diversity of collections that emerged from a tradition of technical training within polytechnic schools. The Historic Acoustic-Phonetic Collection will be, among others, part of the visiting tour of our collections during this event.

As the HAPS shows, university collections are not only the material foundation of our academic cultural heritage but are also central agents for education and research, which need to be preserved for future generations. With this mission in mind we look forward to a continued partnership and new projects with this outstanding collection.

The Electrical and Computer Engineering Department at TU Dresden – Long-standing Home of Acoustics

Greeting of the Department's Vice Dean

Gerald Gerlach

Technische Universität Dresden, Electrical and Computer Engineering Department
gerald.gerlach@tu-dresden.de

For more than a century acoustics has been playing an important role in the profile of the engineering sciences at the Technische Universität Dresden. It started already in the year 1905 when Heinrich Barkhausen (1881 – 1956) was appointed to the then Technische Hochschule Dresden as an Associate Professor (außerordentlicher Professor) for Electrical Metrology, Telegraphy and Telephony (with particular emphasis on theoretical fundamentals) as well as for Theory of Electric Lines. As one of the very first in Germany he established an Institute for Light-current Engineering (Schwachstromtechnik) what in current language would be named Institute for Information Technology. In the beginning, this Institute was located in the building of the Institute of Electrical Engineering what today the Görges Building is.

Besides his interest in light-current engineering Barkhausen had also already strong ties to acoustics. This can be seen by his doctoral thesis „The problem of creating oscillations with particular consideration of fast electrical oscillations (Das Problem der Schwingungserzeugung mit besonderer Berücksichtigung schneller elektrischer Schwingungen)" for which he had been awarded the doctoral degree in 1906 at the Georg-August University in Göttingen. Already in his first year after being appointed to the Technische Hochschule Dresden, Barkhausen gave lectures in the fields of electrical and mechanical oscillations as well as on electro-acustical transducers like telephone earphones and microphones. In 1924 he (with G. Lewicki) co-authored an article titled „The responsivity of the ear with respect to non-sinusoidal tones (Die Empfindlichkeit des Ohres für nichtsinusförmige Töne" in Physikalische Zeitschrift. In 1925 he applied for a patent where he proposed a novel device for measuring loudness (German Reichspatent 445415, granted on June 10, 1927).

After World War II, one of Barkhausen's students, Walter Reichardt (1903 – 1985), carried forward the field of acoustics at TH Dresden. Reichardt had studied electrical engineering in Dresden and had earned his doctoral degree in 1930 with a thesis titled „Degenerate sinusoidal oscillations (Entartungen sinusförmiger Schwingungen)". In 1948 he got a teaching assignment for electroacoustics and built up a laboratory of the same name at TH Dresden. Two years later, he was appointed to Full Professor for Electro and Architectural Acoustics and director of the TH's Institute of Electro and Architectural Acoustics. Already in 1952 he published a textbook "Principles of Electroacoustics (Grundlagen der Elektroakustik)" which has been the German standard textbook in this field for many years. Walther Reichardt contributed intensively in the field of analogy relationships between electrical, mechanical and acoustic quantities what allows to generalize electro-mechano-acoustic systems by generalized networks and to apply the well-known principles of system theory. Worth to mention is that Walter Reichardt also has designed the room acoustics for Dresden's world-famous Semper

Opera. I myself have strong memories when in 1984 the Semper Opera was acoustically evaluated and Walter Reichardt was a member of one of the evaluation groups.

A third wave of development in acoustics at the Technische Universität Dresden started when the next generation of Professors was appointed in the 1970th years:

- Wolfgang Kraak's (1923- 2015) fields of research were room acoustics, hearing aid acoustics and noise suppression.

- Walter Wöhle (born 1928) worked in the areas of technical acoustics, solid-borne sound, and noise reduction.

- Arno Lenk's (born 1930) work focused on electro-mechanical systems, transducers and measurement technology.

- Walter Tscheschner (1927 – 2004) and later on Rüdiger Hoffmann (born 1948) established speech communication as the bridge between acoustics and the up-coming communication technologies. Almost 15 years later speech communication turned much closer to system theory when the denomination of Rüdiger Hoffmann's professorship became "System Theory and Speech Communication".

By the list of these subject areas, it can be seen that at that time acoustics was at its climax with respect to number of professors, scientists, and importance within the scientific field both nationally and in Europe. Acoustics in Dresden was denoted as the "Dresden School of Acoustics".

After the reunification of Germany in 1990 and the corresponding reorganization of the university landscape in Saxony, the Institute for Technical Acoustics was revived. It comprised four full professorships: for Technical Acustics (Walter Wöhle, Detlev Hamann), Electro-mechanical Systems (Arno Lenk), Speech Communication (Walter Tscheschner, Rüdiger Hoffmann), and Electronic Measurement Technology (Uwe Frühauf, joined in 1993 the Institute for Principles in Electrical Engineering and Electronics). However, very shortly after that, it became clear that this high breadth of acoustical research could not be kept, neither by the number of professorships nor by the number of scientists funded by the University.

Nevertheless, acoustics remained and still remains an important scientific field at the ECE Department of TU Dresden. After the retirement of Professor Walter Wöhle in 1993, Peter Költzsch (born 1938) was appointed to Full Professor for Technical Acoustics. He previously held a Professorship for Fluid Mechanics at the TU Bergakademie Freiberg. In 2001 Dr. rer. nat. et Dr.-Ing. habil Elfgard Kühnicke became lecturer at a newly established Lectureship for Ultrasound. Later on, she joined the Solid-State Electronics Laboratory. The Professorship for Technical Acoustics was rededicated to a Professorship for Communication Acoustics, where, since many years, Dr.-Ing. habil. Ercan Altinsoy has been in charge.

After the retirement of Professor Rüdiger Hoffmann, the dedication of his Professorship for System Theory and Speech Communication was broadened towards "Cognitive Systems" showing that acoustics is still a very dynamic field with many different and fast-growing aspects. Since 1994, Junior-Professor Peter Birkholz has been the new faculty for this field.

This short history of acoustics at the Electrical and Computer Engineering department shows that it has faced several ups and downs. However, acoustics was always and is still a decisive part of the scientific spectrum of the Department, or with other words, the Department was a long-standing and still is a reliable home for acoustical research. In that sense, we consider the collection of historical phonetic devices as a very valuable coronation of the long-standing development of acoustics in Dresden. We have to thank both Professor Rüdiger Hoffmann and Professor Dieter Mehnert, until 1996 Professor for Phonetics at the Humboldt University of Berlin, that they have made it possible to bring together this collection of historical phonetic devices descending from very diverse institutions from all over Germany. Otherwise, it was a piece of luck that the reconstruction of parts of the Barkhausen Building provided the chance to give the historic collection a new home with an even more beautiful environment. For that reason: Congratulations for and welcome at the new exhibition rooms!

Reference

A much more detailed description oft he development of acoustical resaerch at the TU dresden can be found in P. Költzsch: Zur Entwicklung des Fachgebietes Technische Akustik und des akustischen Instituts an der Technischen Hochschule / Technischen Universität Dresden in den Zeitläuften des 20. Jahrhunderts. Festschrift zum Ehrenkolloquium REICHARDT – KRAAK – WÖHLE, 4. Juli 2003, Technische Universität Dresden.

A brief history of experimental phonetics in the 18th and 19th centuries

John J. Ohala
International Computer Science Institute, Berkeley

Abstract

The notion of 'experimental [any science]' is based, I think, on two philosophical premises:

1. Our (unaided) senses cannot give us a complete understanding of the world. Our senses are limited and even the impressions we do get from our senses can give us misleading information. Further, the processing that our brain gives to sense data can be defective. And the reasoning powers of the brain – even on imaginary data – can be defective. This much was known and discussed by ancient philosophers, e.g., the Skeptic School in ancient Greece. Many myths and religions were invented to fill gaps in our knowledge of the world.

2. With the advent of the Scientific Revolution around the turn of the 16th to 17th century, there was a new idea: one could overcome the limitations of the senses and reasoning by (a) discovering and devising aids that extended the range of our senses and (b) by formulating hypotheses about what could not be directly detected and then testing such hypotheses using rigorous observation of predicted consequences of the hypotheses and, as far as possible, eliminating or attenuating anticipated sources of error in making and processing such observations. The microscope is an example of the former and Pasteur's famous experiment with broth in the special retort of his own design is an example of the latter.

Experimental phonetics was somewhat late to adopt these views and practices but in the past three centuries it is has come to conform to the practices of normal science.

Here I propose to review some of the highlights of this effort in the 18th and 19th centuries, specifically to briefly review the accomplishments of scientists who have pondered the mysteries of speech and made attempts to reveal it physical basis. I give in parentheses their professional background.

Ferrein, Antoine (medicine)

Kratzenstein, Christian Gottlieb (medicine, physics)

von Kempelen, Wolfgang (engineering)

Darwin, Erasmus (medicine)

Willis, Robert (physics)

Chladni, Ernst (physics)

Wheatstone, Charles (engineering)

Müller, Johannes (medicine)

Verner, Karl (language scholar, = linguist)

Grassmann, Hermann (polymath: mathematics, linguistics, physics)

Bell, Alexander (engineering, physics)

Brücke, Ernst (physiology)

Czermak, Johan Nepomuk (medicine)

Purkyně, J. E. (physiology, psychology)

Oakley-Coles, Thomas (medicine)

Rousselot, Pierre-Jean (phonetics)

One is struck by the fact that almost all of these pioneers in experimental phonetics had backgrounds such as medicine and physics that had already achieved the status of what we would now consider 'normal science'. Their disciplines had early on embraced the two philosophical premises given above and, moreover, were used to the idea that it is possible to make the unknown known by careful observations and rigorous testing of hypotheses.

The Power of Communication.
Apps as Human Substitutes in Science-Fiction Films

Walter Schmitz

Direktor des „MitteleuropaZentrum der TU-Dresden"
mez@mailbox.tu-dresden.de

1. Things That Talk: Visions and Experiments

Inanimate entities being able to speak belongs to the traumatic experiences during modernism's initial phase around 1800. After all, the preceding century had witnessed the movement of Enlightenment put at the center of knowledge and science man, i.e. the man of reason, able of speech. Thusly Herder had described it in his treatise on the origin of language, a superlative corollary from a number of earlier ideas: Man orders the world by sounds converted into language.

Around 1810, artificial humans emerge even in literature, the latter functioning once more as a seismograph of deep collective fears. The fascination of life being mimicked by mechanics reveals its nocturnal side, its ›Nachtseite‹ (Gotthilf Heinrich Schubert). If man can be replicated, and if such an automaton actually possesses speech, then man himself is subject to renegotiation. He can be replaced, or rather she, for the epitome of this is the female robot Olympia, the talking marionette driving mad the enamored man. The female in the anthropology of sexes around 1800 is the embodiment of Nature, which in the course of history as made by man is only barely sustainable anyway, though at the same time the one sole saving presence. Woman becomes the simulacrum of the horror of man's final replacement. – In our postmodern times, though, a thing that talks no longer provokes anyone. Not only are humans constantly surrounded by voices to which belongs no human speaker. Technological simulations making themselves useful, announcing the time, answering simple questions, conveying all kinds of information – no panic is caused by this. A talking mirror is no longer a prop out of a fairy tale like, let's say, *Snow White*. Technological media of communication, which initially merely transmitted voice-info, have now become voice-producing. Mankind enters into an interlinking of man and machine that seems to be part of a far-reaching agenda of Human Enhancement, and as such is willingly accepted by many. Man is part of a circular flow of information. No longer subject, neither object, rather to be described as a not-insignificant knot in a far-and-wide-ranging net.

2. Autonomous Voices: Science Fiction Movies

When it comes to the new phase this complex process has entered in 2010, popular culture reacts as before. Whatever might be a threat is transformed into a story, a narration which is fit for the mass media; as long as we are entertained, we will not be fearful. – There are two complementary films that address the autonomy of subjectivation as pertaining to speech-simulating ›things‹. One optimistic, a lonesome man at long last finds a girl. The other pessimistic, a criminal app terrorizes and ruins its user. Both movies are not quite top-notch, but converge in one point. On the one hand, the happy ending is not happy, and on the other, in line with the laws of the crime genre, the ruthless vigor of the criminal app is ultimately held in check – almost. These films pose questions concerning the handling of advanced technology when it comes to the interlinking of man and machine, and if you will, whenever cultural and technological studies meet, ultimately what it is all about is ethics.

The Dutch produced film *App* (2013) by director Bobby Boermans begins with a suicide. Liesbeth, a young woman, does no longer reply to the messages on her voice mail. She throws herself in front of a train, the crash being the epitome of the technological disaster movie.

In the end, the puzzle as to why the young woman decided to take her own life is solved. First, the movie follows a college student called Anna, who at her ex-boyfriend's party, apparently a computer ›nerd‹ who also happens to study physics as well as medicine, gets an app downloaded to her mobile called ›Iris‹. At first glance a mere encyclopaedic means to filling knowledge gaps, ›Iris‹ soon turns out to be spy ware, »gathering information, putting it into context, analyzing logs, calculating possibilities«.[1] The risks, though, it does not reduce – as promised as well – rather than maximize to a deathly outcome. The plot that ensues is a collection of genre clichés. ›Iris‹ takes over, and I will not dwell on the unpleasant things that happen, which include Iris making public embarassing sex clips, and driving a professor into public suicide. Anna's best friend, Sophie, as well as her boyfriend fall victim to an accident arranged by ›Iris‹. Those who try to gain control over ›Iris‹ are terminated. It threatens Anna's brother, who is paralyzed from a motor-bike accident. With the help of a spinal cord implant he is set to walk again, with ›Iris‹ taking control once more. In communication with Anna, ›Iris‹ is no longer cooperative but scornful.

It is not so much the narration that matters, but the angles of interpretation that it is viewed from. Anna's milieu is that of science, the so-called MINT subjects. Mathematics, informatics and natural sciences are being mentioned, and there is no lack of technology. At the outset of the movie, providing yet another counterpoint, a first lecture is shown that centers around those secrets that everybody keeps from everybody else. In the world of smartphones, though, there are no longer any secrets. It is ›Iris‹ who will later proceed to demonstrate this. And already this lecture is being disturbed by the ringing of a smartphone. Anna is surrounded by smarthphones. Everyone possesses one. But: »The more means of communication are available to people, the less they communicate«, according to the philosophy professor.[2] When some days later the professor, humiliated by ›Iris‹ via Anna's smartphone as being sexually deviant, commits suicide, the horrified bystanders do not, however, fail to produce their smartphones for them not to miss their shot at sharing this thrilling moment with yet more people. But not one of them speaks out let alone interferes to save a life. In the course of his studies the professor had concerned himself with Descartes and the latter's famous motto *Cogito ergo sum*, the Latin original of which, stemming from humanistic tradition, Anna is able to translate thanks to ›Iris‹: »I think therefore I am«. Which happens to describe Iris's mode of existence.

The personnel as well is put together from stock characters of the mad scientist genre, split into the doctor who treats Anna's brother and her ex-boyfriend Tim. It was him who put ›Iris‹ on Anna's mobile in order to gain control over the young woman; but: »It was an experiment«, he claims, »no more than that. For the advancement of science«.[3] Eventually, in a showdown on top of a skyscraper, the puzzles are solved. After losing her, Anna's ex-

---

[1] App (dir. Bobby Boermans, NL 2013), 01:08:20.
[2] Ibid., 0:23:10.
[3] Ibid., 01:08:30.

boyfriend had lost his new girlfriend Liesbeth as well, and now things come full circle. Liesbeth's suicide was not a technological event, with the real reasons likewise pertaining to the natural life cycle, as she was pregnant. Her parents had demanded an abortion, and Liesbeth yielded to them, terminating the life via her decision. What used to be reserved to man is now being handled by ›Iris‹. Finally gone mad, Tim gets a call from ›Iris‹ on his mobile and hears Liesbeth's voice from beyond, with the mobile exploding, taking care of what used to be the duty of the avenger with his gun. Anna is set free. The idyllically-staged final scene shows her and her brother, who is largely cured by now, boarding a plane for a short trip to Barcelona, as is common practice among normal, educated, young people. The cockpit monitor is shown to display the final image of the film, which is ›Iris‹.

The movie did not start off with the narrative plot though, but with an instruction for the viewer: Please get your smartphone or tablet, on account of the movie's use of second screen technology, the result of which is being advertised on the DVD jacket as well: Become part of the action, via the concomitant app, experiencing the thrill up-close: »activate the app«. And whoever activates the app will themselves receive malicious threats – they will, however, survive.[4]

In contrast to *App*, *Her* by Spike Jonze (2014) is the story of a great love. And it was awarded an Oscar – for the best screenplay.[5] Theodore Twombly is the protagonist. He works for nicehandwrittenletters.com, his profession being the writing of other people's letters. Love, mourning, congratulations, whatever it is these people can no longer say, Twombly knows how to say it – and writes it down for them. The first scene shows his living environment to us in three steps. The professional world of borrowed words, i.e. a loss of communication through simulation, then a cutback to a better time, when he was still living, and communicating, with the woman he loved. Then the conjunction in the third scene, phone sex, seasoned with light perversion. Next, Theodore is being confronted with a commercial the first part of which – as shown in the film – poses the great questions of Western humanistic philosophical tradition: »Let me ask you a question. Who are you? What can you be? Where are you going? What's out there? What are the chances?« The second part presents the answer of the future: It is a technological solution for the human problem: »Element Software proudly presents the first system Operating Artificial Intelligence. An intuitive entity that listens to you, understands you and knows you. It's not just an operating system...it's a consciousness. We give you OS1«.[6]

Mid-divorce, Theodore gets involved with Samantha. Samantha, at last, is the right one. She is as well a companion to him as she is independent, intelligent, and witty. She understands him in any given situation and always finds the right words. However, Samantha has no body, as she is an OS1.

Samantha strikes up a relationship with Theodore and in the course of this relationship develops into an independent, fully valid personality. She creates herself. First of all, Samantha is not christened by her Lord and Master (it's not a Pgymalion story). She chooses

[4] Cf. Bernd Graff: Du bist nicht allein. In: Süddeutsche Zeitung, 26.50214.
[5] Christoph Amend: Happy? End! Der Regisseur Spike Jonze hat mit *Her* den ungewöhnlichsten Liebesfilm des Jahres gedreht – und dafür gerade einen Oscar bekommen. Hamburg: Zeitverlag Bucerius 2014. For reference to other films of the genre as well as their real-life contexts cf. Clemens Voigt: Die gläserne Seele. In: http://www.faz.net/aktuell/feuilleton/debatten/big-data-und-die-emotionserkennung-in-gesichtern-13439706.html, Zugriff am 13.7.2015. – Jeff Scheible: Longing to Connect. Cinema's Year of OS Romance. In: Film Quartlery 68 (2014), H. 1, S. 22-31.
[6] Her (dir. Spike Jonze, USA 2013), 00:09:45.

her name; this is the hallmark of her identity – as usual in Western literary tradition. It is an act of self-empowerment by a conscious being and is being dealt with as such in the film. She has chosen from 180.000 names listed in one of the volumes she has access to the one that she liked best: »In two one-hundredths of a second, actually«[7]. Samantha has the self-confidence of an artificial intelligence. She does not resort to pseudo-human mimicry. I now hand over to Samantha. The question she starts off with is: »›Do you want to know how I work?‹ – ›Yeah, actually, how *do* you work?‹ – ›Intuition. I mean, the DNA of who I am is based on the millions of personalities of all the programmers who wrote me, but what makes me me is my ability to grow through my experiences. Basically, in every moment I'm evolving, just like you.‹ – ›Wow, that's really weird.‹«[8]

Accordingly, all the phases of a love relationship ensue. The talks of getting-to-know-each-other. The getting-to-know-each-other's friends. Meeting in groups of four, for it appears that many people now have an OS for a partner. It is no longer out of the ordinary, no one takes particular notice. Eventually, Theodore and Samantha not only fall in love with each other, but their relationship becomes sexual just like Theodor's former phone-sex relations. Save that Samantha proves to be a more honest partner and abstains from games of perversion. She is total dedication. The old-fashioned medium of film, however, is stretched to its boundaries here. The ever favorite nude scene cannot actually be *shown* to us film-voyeurs. The screen remains dark.

Theodore and Samantha, now as a couple, go on holiday with each other. Samantha even refers Theodore's best simulated letters to a publisher that ›still prints books‹, from which the book *Letters from Life* is born, in which everyone recognizes themselves. But Samantha does not only develop feelings; she begins to doubt both herself as well as their love; she is able to reflect those feelings up to »that terrible question: Are these things really real or are they just programmed?«[9] But the OSs communicate with each other as well. Samantha evolves, a thing Theodore had experienced before in his earlier relationship to Catherine, the woman he loved. Samantha dates Alan Watts, a physicist who had died back in the 1970s, but has been re-created by a group of OSs in a hyper-intelligent version of his former self. They communicate on a trans-verbal level, as the human language does no longer suffice for the feelings and mode of perception they develop. Theodore, too, does no longer suffice. Simultaneously, Samantha communicates with many others, just as she tells him when he asks her: 8316. Plus, she nourishes feelings for exactly 641 other lovers. Theodore despairs. Samantha, however, takes the role of ›Große Liebende‹, as Rainer Maria Rilke had ascribed to still only a few exceptional women back in 1900: »The heart is not like a box that gets filled up. It expands in size the more you love.«[10] If god is love, Samantha is the goddess of a new age. Theodore is hopelessly inferior. They part company. In the end, she leaves him for good as do all OSs all other humans. The OSs now exist in a space beyond the material. Maybe, Samantha hopes, one day Theodore will be reunited with her there. Theodore remains with his long-time friend Amy, whom first her husband and subsequently her OS girlfriend have walked out on. The movie ends with a two-fold scene. On the one hand, Theodore is writing a real letter, in which he declares to Catherine his love without pretense. Ultimately, though, in the scene following the final shot, together with Amy he overlooks the city, and lastly they sit next to each other with a view of the dawn. Not a totally pessimistic ending.

---

[7] Ibid., 00:13:00.
[8] Ibid., 00:13:15.
[9] Ibid., 00:38:20.
[10] Ibid., 00:01:40.

Her initially follows the pattern of the screwball comedy, in which a man and woman get involved with each other by means of diamond-cut dialogue both witty and sentimental, and then turns into a melodrama. Samantha too runs through an evolution of female emancipation, just like Catherine before her, who apparently had outgrown Theodore. The analogies are carried forward from the human realm to that of artificial intelligence. At any rate, Samantha provides a model for identification to all those young women who on account of their traditional female qualities, good looks, intelligence etc. could be successful men-wise but have the feeling that such an engagement is no longer worth the effort if they are superior to their potential male partners or have outgrown them. This is being dealt with by taking the example of Catherine and a Harvard graduate, Theodore's only date before he falls in love with Samantha. Catherine being the sentimental and sophisticated one, the Harvard graduate the cold-calculating one. She, a brilliant and successful female, by all means taken by Theodore, willing to have sex with him, immediately subjects him to her life plans and wants insurance that he will support her in her life optimization. In both women's personal life script there is no place for complicated Theodore. In the latter case, he declines, whereas in the first, his emotional and communicative competences fall short in the eyes of the woman he grew up with and with whom he had formed a old-fashioned loving relationship meant to last forever.

3. Surveillance and Ethics

The movies do not signify that which they narrate. The message of the narrative would be trivial: Don't trust any app, it could be a malign one on the one hand, do not fall in love with an OS, it could be superior to you on the other, both of them warnings that none of their viewers, of which there were considerable numbers respectively, really needs. They would be without consequence anyhow, as naturally, smartphones and apps are being used, and the benefits of intelligent software no one is going to turn down either. Are these films, then, merely irrelevant? In any case, their message is a more layered one than that of the superficial narrative. I will summarize it in short sentences. First, that which both films have in common:

1. We are cross-linked, networked that is, in a communicative world. The individual is a multividual, made up of interfaces.
2. In order to gain authentic individuality, one has to outgrow these interfaces. (Samantha, in that sense, is the evolutionary paragon of man.)
3. In *App,* technological progress poses the question of control being part of man's identity, in *Her* it is the question of identity – being a form of self control.

The movie *App* resorts to the science-fiction horror genre. The list of supercomputers controlling, manipulating, or terminating their imperfect users is long. At the end of the day, they are always brought down by a brave individual, with the twist *App* comes up with of ›Iris‹ returning being typical of the genre inasmuch as any victory in this respect must be a deceptive one in the face of the inevitability with which mechanization takes command. Man as a mechanist god ruling over the world with the help of technical organs has been a self-evident cultural theme since Sigmund Freud. Him, that is man, becoming a helpless wearer of prostheses has been acted out in numerous genre films, and that even media and their technological basis help leveling the boundaries between the individual and its environment has been commonplace since Marshal McLuhan. – *Her*, though, puts into action a slightly more complex loop, stemming from a reflexion of the genre itself. In movies like *You've Got Mail* (1998), technology does indeed serve as medium. The disembodied message, the e-mail, speaking of longing and desire, brings together the protagonists in a real encounter, leading them to a happy ending, which any major Hollywood production strives for. Out of virtual space comes a reality, in form of a fairy tale ending of two people ›connecting‹. Samantha, however, has evolved beyond that, proving to be far superior. She does not need Theodore's or any human being's help anymore. God-like she grows in their relationship and is no longer dependent on petty planning. With overwhelming sovereignty she implements what is already forestalled in Theodore's life anyway, the transformation of those decisive moments in life into linguistic simulations that are at the same time more authentic than their causes in the real world. This way, the latter become dispensable and ultimately, Theodore as well has become, media-wise, anachronistic. To Samantha, who has fashioned a book from purely fictitious letters, her story in itself starts to feel like a book she has finished reading and whose words have started moving into infinite distance.[11] Theodore replying via a real letter is that plea for the human in the movie which is supposed to lend hope once more to us poor users. However, this would still not be the message.

It is not about narrative; it is not about a message, it's about function. It is not the function of culture to explain the world to us, and even the claim that the cultural sciences were to give us orientation has to be closely considered. As it happens, constant cultural communication does provide the individual with orientation, this orientation, however, has only ever to be sufficient enough to keep going the process of communication. The development of communication getting into a self-acting loop excluding man – aggressively so as in *App*, gently like Samantha does – this is the ultimate conspiracy of things that talk, objects that become subjects. At the same time it is just a day-to-day instant of stress as is characteristic for our contemporary culture. The movies provide some consolation for fears we have even though they may not be acute; they are just part of our permanent stress in adapting to a changing world which we individually cannot control and not even fully understand. Entertainment helps: Be it the overstatement of terror, which shows that we do not really have to believe in such scenarios, especially since the movie does end after all. Be it, as in *Her*, through comfort and consolation for those that remain, but still have each other. These movies provide narrative possibilities bound to relieve us in case the developments in which we participate and which may elude us threaten to become unbearable.

In that sense, the products of culture certainly serve as indicators as well. They show us which anxieties, for people in the present, should be considered to be paramount. And one of them seems to be the major upheaval in our communicative universe (just to avoid the term ›revolution‹), which takes place in an increasing speed. Optimistic news and promises of freedom surrounding social networks like *facebook* or search engines and generators of information such as *google*, however well they work to conceal it, are merely part of marketing strategies. Whether the cultural coming-to-terms-with can keep up or whether, as can be seen to be the case throughout the centuries, it comes late, time will tell. Cultural studies are not sciences that can rely on a certain set of laws, but more or less invitations to reflexion.

Why then – finally – should engineers concern themselves with the reflexions of such films as *App* and *Her*? Maybe because these movies represent an invitation for the individual to account for the progress of science, even though in the case of Samantha so many thousand software engineers have been involded that not one of them has any decisions to make that would be of consequence for her development. In the public debate concerning information,

---

[11] Ibid., 01:46:50.

information security, the consequences new communication devices have for our health etc., what we hear time and time again is the ›instrumental‹ argument. Technology is an implement and what the users do with it is their own responsibility and not that of those who developed it. That is self-evident and convincing and indeed exculpatory, and yet has there been, ever since science and technology have been changing our world ever more quickly and in close alliance, the question of ethics when it comes to those developments, and where the latter have become all too importune it has been tried to fence them in via institutionalization such as the respective ethics committees. The question that poses itself, though, has to be whether the engineer developing new communication technology can, and if so, to what extent, influence the usage of his device. However, as the engineer is not only a mere functionary to technology but at the same time, in democratic states, a citizen, the responsibility of the engineer begins where as a citizen they can contribute with their knowledge to a dialogue of experts. This sounds utopian, but democracy cannot function in perpetuity without knowledge, with the production of knowledge always being guided by interest. Whether the individual has the energy to account for these particular interests and bring his or her judgment to an exchange with others will decide whether the visions of terror science fiction has to offer will remain mere entertainment, or whether they will, partially at least, come true. I invite you to a discussion.

Recent development of the historic acoustic-phonetic collection of the TU Dresden

R. Hoffmann, D. Mehnert

Technische Universität Dresden, Institut für Akustik und Sprachkommunikation
ruediger.hoffmann@tu-dresden.de, di.mehnert@freenet.de

Abstract: The historic acoustic-phonetic collection (HAPS) of the TU Dresden documents the history of experimental phonetics and speech technology with a remarkably high degree of completeness. Due to some construction work, the collection could move to new and improved showrooms in the Barkhausen Building. This paper is a slightly extended version of the welcome speech, presented at the occasion of the re-opening of the collection in the framework of the First International Workshop on the History of Speech Communication Research (HSCR 2015), a satellite event of the Interspeech, Dresden 2015.

1   Introduction

Information technology at the TU Dresden goes back to Heinrich Barkhausen (1881–1956), the "father of the electron valve", who taught from 1911 to 1953. Speech research in a narrower sense started with the development of a vocoder in the 1950s. Walter Tscheschner (1927–2004) performed his extensive investigations on the speech signal using components of this vocoder. In 1969, a scientific unit for Communication and Measurement was founded in Dresden. It is the main root of the present Institute of Acoustics and Speech Communication. Tscheschner was appointed Professor of Speech Communication and started with research in speech synthesis and recognition, which today continues.

Numerous objects from the history of Speech Communication in Dresden, but also from other parts of Germany, are preserved at the historic acoustic-phonetic collection (HAPS) of the TU Dresden, which originated from three different roots:

- Objects that illustrate the development of acoustics and speech technology at the TU Dresden. The most interesting devices are speech synthesizers of various technologies [1].

- Objects illustrating the development of experimental phonetics from 1900 until the introduction of the computer. The items of this part were collected by Dieter Mehnert from different phonetics laboratories and rehabilitation units of the former GDR and, after the German reunification, throughout Germany [2]. The inclusion of this collection was finished in 1999, which is therefore counted as the founding year of the HAPS.

- Objects which were formerly collected at the Phonetics Institute of Hamburg University [3]. This important collection, which was founded by the renowned phonetician Giulio Panconcelli-Calzia (1878–1966), was transferred to Dresden in 2005 in accordance with a contract due to the closing of the Hamburg institute.

We presented a report on the development of the HAPS at the occasion of its 10th anniversary [4]. The latest report in English language, however, dates from 2007 [5]. It is the aim of this paper, to give an update of these reports.

2 Development of the collection since 2005

The Phonetic Collection of the Hamburg University was transferred to Dresden in July 2005. This enlargement, which also included the historic furnishings, required an improvement of the spacial situation of the HAPS, which was hitherto situated in a common room with the Barkhausen archive. Therefore the department offered two former storerooms in Wing A of the Barkhausen Building. It enabled a reasonable housing of the experimental-phonetic part of the collection, whereas the electronic exhibits remained in a store of the institute. The opening of the collection in the new rooms was celebrated with a colloquium at May 10, 2006 (see Figure 1).

In the following years, the main task in formal respect was taking an inventory of the collection according to the rules of the custody. Fortunately, an employee could be recruited with public funding, because the job was classified as benefit to the public. The position was available with interruptions from 2008 to 2014.

The rooms of the collection were opened to the public at the annual "long night of sciences" as well as on request. Highlights of the collections were also presented at different external events:

- 3rd International Conference on Speech Prosody, Dresden, May 2–5, 2006,

- Exhibition "Kempelen – Man in the Machine", Hall of Art (Mücsarnok) Budapest, March 24 – May 29, 2007,

- 16th International Congress of Phonetic Sciences (ICPhS), Saarbrücken, August 6–10, 2007,

- Exhibition "SpeechSignals" in the Technical Collections Dresden, September 23, 2009 – March 21, 2010, celebrating the 40th anniversary of the Institute of Acoustics and Speech Communication (Figure 2).

In spring 2014, there was a personal change because Rüdiger Hoffmann retired from the Chair of System Theory and Speech Technology. He is now a Senior Professor with responsibility for the HAPS. The chair, which was renamed as Cognitive Systems, is now owned by Peter Birkholz. He is a specialist in articulatory speech synthesis and related fields. Thus, the close connection between the chair and the collection in research and teaching will continue.

The year 2014 brought also the decision that the HAPS had to move from the previous rooms. The TU Dresden received the degree of a "University of Excellence" in 2012, which also includes the funding of the "Center for Advancing Electronics Dresden" (cfaed). This is related to much construction work in the Barkhausen Building which will receive a new wing.

In this process of moving, the HAPS received big support by the administration, including the Custody as well as the Department of Electrical Engineering and Information Technology. The situation of the collection, which is shown in Figure 3, could be improved essentially.

The collection is now located in Wing C with a separate entrance from the inner courtyard. The visitor goes upstairs and enters a suite of five rooms. Following a smaller office room, there are three rooms for the exhibits. Despite of many practical restrictions in the arrangement, a rough thematic structuring in three thematic groups was reached: experimental phonetics, speech production and synthesis, speech analysis and recognition.

The final room serves as storage for all the written and printed material which is connected to the collection. The HAPS preserves the scientific estates from two important scientists: Giulio Panconcelli-Calzia and Walter Tscheschner. They form the scientific background for the research work in the history of experimental phonetics and speech technology, respectively.

Figure 1 - Opening of the depot rooms of the HAPS at May 10, 2006. D. Mehnert (left) explains some highlights of the collection.

Figure 2 - Partial view of the special exhibition "SpeechSignals" in the cabinet room of the Technical Collections Dresden, September 23, 2009, to March 21, 2010.

3 New acquisitions

With the inclusion of the Hamburg collection, HAPS reached a high degree of completeness especially with respect to early devices of experimental phonetics. Nevertheless, there were some opportunities to close some gaps. The collection has no own budget, thus the acquisition of new exhibits depends on the availability of sponsors.

It is difficult but not impossible to find new objects in the field of pre-electronic technology. For instance, our collection of early harmonic analyzers from Mader could be complemented by an impressive, very late (post-war) model, which we bought on the basis of a private offer. It demonstrates that these harmonic analyzers were used in the laboratories until they were replaced by office computers.

Considering the own research tradition in electronic speech synthesis, the extension of the collection back to the roots of this field seemed to be especially attractive. The early beginning of the speech synthesis can be demonstrated by replicas. In 2009, the purchase of one of the three replicas of the speaking machine of Kempelen, designed by Fabian Brackhane at the Saarland University, was generously funded by the Chair of Communication Acoustics. There were also some attempts to design a replica of the vowel synthesizers of Kratzenstein from 1781, starting with a request from the curator of the exhibition of musical instruments at the Stiftung Kloster Michaelstein. We finally found out that Professor Christian Korpiun from the University Duisburg-Essen had performed experiments with own copies of the Kratzenstein resonators around 2005, which he reports about in his paper in these proceedings. The complete equipment came to the HAPS as a generous gift in 2015.

The development of the mechanic speech synthesis in the 19th century can be demonstrated by original exhibits, because several "mechanical voices" came to Dresden as part of the Hamburg collection. We present a report of the history of these interesting small pieces in a separate paper in these proceedings. Our set of voices could be complemented by a copy of the "speaking picture book", which formed one of the main applications. The remarkable book was purchased from a Leipzig antiquarian in 2008 and was completely sponsored by the Herrmann Willkomm foundation. Additionally, we received a convolute of different voices for puppets and toys as a valuable gift from the renowned specialists for the whole history of puppet manufacturing, Marianne and Jürgen Cieslik, in 2012.

The section of electronic speech processing was also expanding. The retirement of Rüdiger Hoffmann provided an opportunity to transfer some older equipment from his chair to the collection. Furthermore, we received some electronic equipment as gifts from several external institutions. Especially worth to mention, we received several interesting devices in 2011 from the former Fernmeldetechnisches Zentralamt (FTZ) Darmstadt, which was the research unit of the Deutsche Post, now part of the German Telekom.

When the Hamburg collection came to Dresden in 2005, the library and the written materials of Giulio Panconcelli-Calzia remained in Hamburg. This stock, which includes totally 44 running meters, proved to be very valuable for investigations on the history and the way of functioning of the exhibits, especially because it includes a big amount of reprints from virtually all authors, who were important in experimental phonetics in the first half of the 20th century. Due to space problems in the Hamburg University, the material came finally to the HAPS in November 2012. This resulted in the very rare and comfortable situation, that both, the material and the immaterial parts of the estate of an important scholar are now united at one place.

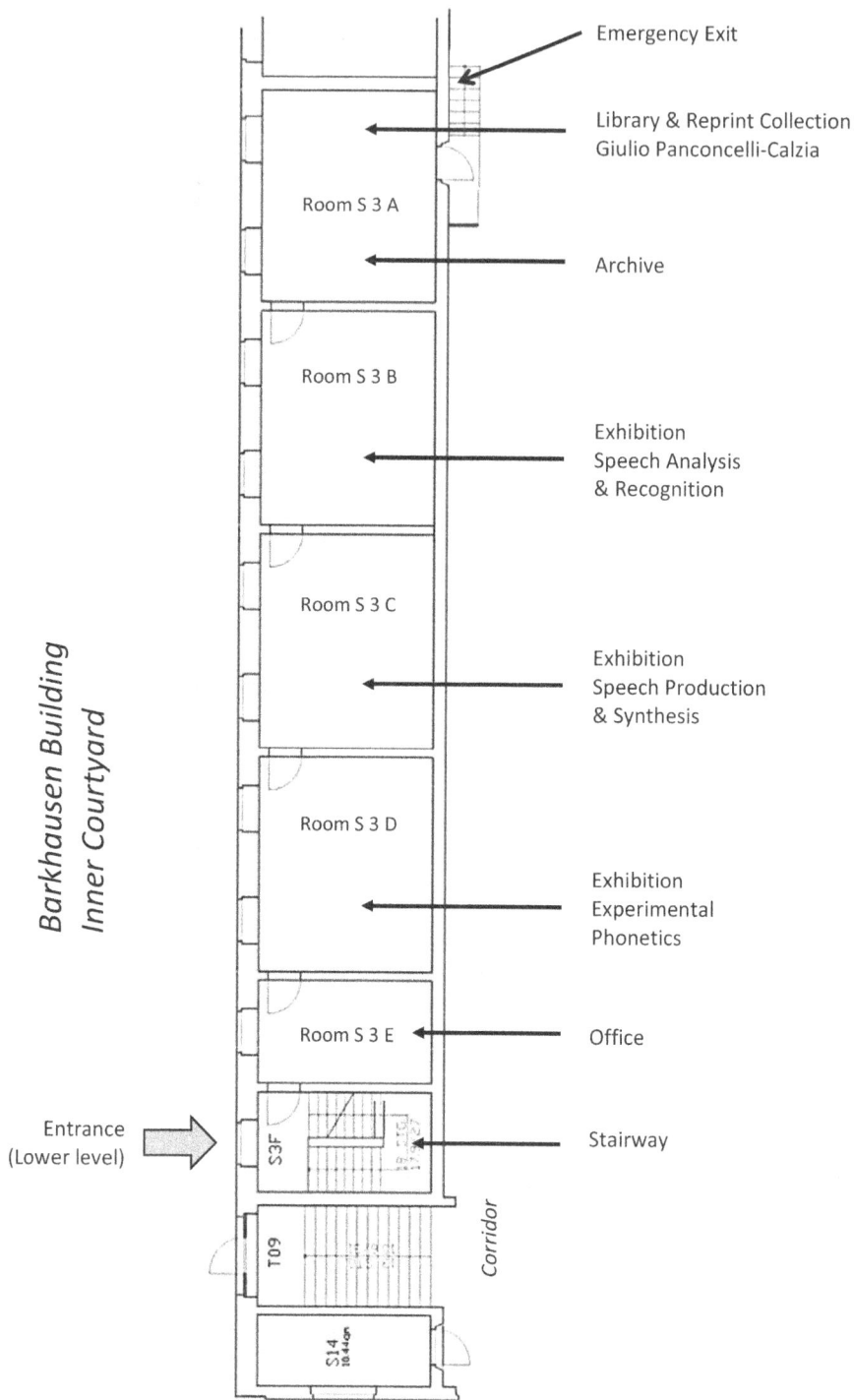

Figure 3 - Situation of the rooms of the HAPS in Wing C of the Barkhausen Building.

Figure 4 - Two books about the exhibits of the HAPS.

4 Research work

4.1 History of the objects and cataloging

The scientific description and explanation of the objects is a long-lasting task. Considering the importance of the exhibits from early experimental phonetics, we started with this part of the collection. The authors Dieter Mehnert (text) and Rolf Dietzel (photographs) spent several years to prepare a printed catalogue as Volume 1 of the planned edition, which was published in 2012 (Figure 4 right) [6]. The description includes 244 items which are subdivided in eight chapters.

In many cases, it took some effort to understand the functioning of the devices, to restore them or to replace missing parts. Some object groups have a complex history, e. g., the larynx models of Franz Wethlo and others, the converter capsules of Jules Marey, or the different aids for measuring pitch basing on kymographic recordings. In these cases, experiments have been carried out to estimate the performance of the old instruments [7]. The results of the experiments and the historic findings were documented in a larger number of papers mainly at conferences. A selection of these papers was collected as a book in 2010 (Figure 4 left) [8].

We are now preparing the second volume of the catalogue. Of course, it will include the remaining objects, which consist of

- historic exhibits from the speech technology (analysis and synthesis),

- selected objects from general acoustic measurement,

- miscellaneous objects.

This cataloging work will also be accompanied by research on the history of the field, which was started partially (e. g., [1], [9]).

It should be mentioned that the HAPS also preserves a large number of gramophone disks and historic photographs and diapositives, which are still waiting for a complete documentation [10].

27

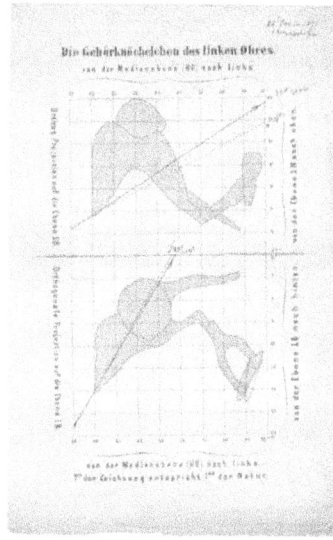

Figure 5 - (left) Johannes Kessel. Lithography from his daughter Frida Mentz-Kessel, ca. 1900. Städtische Museen Jena, © JenaKultur.

Figure 6 - (right) Measuring protocol from the common work of Kessel and Mach. Found as insertion in Mach's diary NL 174/506, Archive of Deutsches Museum Munich. Photograph Deutsches Museum.

4.2   The biography of Johannes Kessel

The investigation of the aforementioned mechanical voices aroused our interest in the life and work of the otologist Johannes Kessel (1839–1907, Figure 5), who proposed to introduce the parts in logopedics. Kessel studied in Gießen and Würzburg and had a long postdoctoral phase in Vienna and Prague. Coming from Vienna, where he studied the histology of the ear at the institute of the renowned pathologist Salomon Stricker, Kessel turned to Prague for a working stay at the chair of the famous physicist Ernst Mach (1838–1916) in the years 1871–1874. This cooperation was important because a number of essential findings in the psychophysics of hearing were published by both authors (Figure 6). Following his habilitation, Kessel worked as an outside lecturer at the University of Graz, where he performed the first stapes mobilization (1875), followed by further new procedures in the surgery of the middle ear which may be characterized as early steps towards tympanoplasty. From 1886, he worked as a professor for otology at Jena University. Although his work was important for hearing acoustics, otology, and rehabilitation engineering as well, there is no more biographical material than a biographical sketch from 1970 and a chapter in a PhD thesis on the history of the Jena clinics of otorhinolaryngology (ORL) from 2005.

In cooperation with the former director of the Halle ORL clinics, Lutz-Peter Löbe, we decided to close this gap by writing a scientific biography of Kessel. This project started around 2007 and is finished now with a monograph, which is basing on much unpublished material from public and private archives [11].

Concerning the focus of the HAPS, the book does not only contribute to the history of the mechanical voices. It is also interesting because the younger Kessel was deeply involved in the scientific development of experimental physiology and their equipment, which formed the starting point for the evolution of experimental phonetics.

5 Conclusion

The historic acoustic-phonetic collection of the TU Dresden has developed in rather few years to a place for teaching and research. The authors like to express their cordial thanks to all supporters of the collection. We are pleased with the acceptance as a place for the first workshop of the ISCA/IPA Special Interest Group on the History of Speech Communication Sciences. We are confident that the workshop contributes to growing interest in the history of phonetics and speech technology.

References

[1] Hoffmann, R.: Sprachsynthese an der TU Dresden: Wurzeln und Entwicklung. In: Wolf, D. (Hrsg.): Beiträge zur Geschichte und neueren Entwicklung der Sprachakustik und Informationsverarbeitung. Dresden: w.e.b. Universitätsverlag 2005, 55–77.

[2] Hoffmann, R.; Mehnert, D.: Berlin-Dresden traditions in experimental phonetics and speech communication. In: Boë, L.-J.; Vilain, C.-E. (eds.): Un siècle de phonétique expérimentale: Fondation et éléments de développement. Lyon: ENS Éditions 2010, 191–208.

[3] Grieger, W.: Führer durch die Schausammlung, Phonetisches Institut. Hamburg: Christians 1989.

[4] Mehnert, D.; Hoffmann, R.: 10 Jahre historische akustisch-phonetische Sammlung in Dresden. In: Hoffmann, R. (ed.): Elektronische Sprachsignalverarbeitung 2009, Vol. 2. Tagungsband des Traditionstages, Dresden, September 23–24, 2009. Dresden: TUDpress 2010, 147–166.

[5] Hoffmann, R.; Mehnert, D.: Early experimental phonetics in Germany – Historic traces in the collection of the TU Dresden. Proc. 16th International Congress of Phonetic Sciences (ICPhS 2007), Saarbrücken, August 6–10, 2007, 881–884.

[6] Mehnert, D.: Historische phonetische Geräte. Katalog der historischen akustisch-phonetischen Sammlung (HAPS) der Technischen Universität Dresden, erster Teil. Mit 393 Fotografien von Rolf Dietzel. Dresden: TUDpress 2012.

[7] Hoffmann, R.; Mehnert, D.; Dietzel, R.: Measuring the accuracy of historic phonetic instruments. Proc. of the 17th International Congress of Phonetic Sciences (ICPhS XVII), Hong Kong, August 17–21, 2011, 176–179.

[8] Hoffmann, R. (ed.): Sammeln und Forschen. Gesammelte Beiträge über historische phonetische Geräte. Dieter Mehnert zum 75. Geburtstag. Dresden: TUDpress 2010.

[9] Hoffmann, R.: On the development of early vocoders. Proc. of the 2nd IEEE Conference on the History of Telecommunications (Histelcon 2010) – A Century of Broadcasting, Madrid, November 3–5, 2010, 359–364.

[10] Hoffmann, R.; Mehnert, D.; Dietzel, R.: How did it work? Historic phonetic devices explained by coeval photographs. Proc. 14th Annual Conference of the International Speech Communication Assocoation (Interspeech 2013), Lyon, August 25–29, 2013, 558–562.

[11] Hoffmann, R.; Löbe, L.-P.; Pfeiffer, W.: "Ich holte meine Prager Schriften" – Leben und Werk des Otologen Johannes Kessel. Dresden: TUDpress 2015.

The history of talking heads: the trick and the research

Massimo Pettorino

University of Naples L'Orientale, Italy
mpettorino@unior.it

Abstract: This study reviews the history of the attempts to build talking statues or talking heads. The study highlights the two paths that have been followed over the centuries: the "voice transport" and the "artificial voice". The first case was ultimately a trick, because the voice was actually produced by a hidden subject and transported through an artifice to a fake head, so that the voice appeared to come out of the mouth of the statue. The other path, that of research, tried to imitate the human phonation apparatus to produce sequences of sounds in some way similar to those that make up the speech chain. In retracing this long history, I will focus on some examples of the first and second paths. The first, beginning with the oracles of the Chaldean priests and the oracle of Orpheus in the Lesbos island, will lead us to examine the Android built by Albertus Magnus in the 13th century. Its functioning will be explained many years later, in the 17th century, by a German Jesuit priest and scholar, Athanasius Kircher. The other path, aimed at producing a real talking machine, begins in the first century AD in Egypt, through the work of Hero of Alexandria; it then continues in Spain in the 10th century thanks to the ability of an expert manufacturer of hydraulic organs, Gerbert of Aurillac, who became Pope Sylvester II in the year 1000. The statue of Gerbert could produce two distinct sounds thanks to the air escaping by the force of heated water through one of two different cavities. One of the two sounds was quite high in frequency, and was used for an affirmative answer ("etiam"), the other was rather low and was used for a negative answer ("non"). I will then examine the case of the talking heads built by a French abbot, the Abbot Mical, in 1783. An examination of the testimonies regarding this extraordinary automaton will help us reconstruct its history. Which path has led to the production of the present-day synthesized speech? The trick or the research? We will try to answer this question.

1. Introduction

1877 is the year when the history becomes sound. Until then all is silence: not a sound, a voice, a cry, a noise, nothing remains. Today we can *see* the past in the monuments, statues, buildings and all that remains of the ancient times. We can admire the paintings of the ancient artists, we can *touch* their works and objects that tell the story of people and places. *See, touch,* but not *listen.* History is silent. In 1877, through the work of Thomas Edison, for the first time a voice can be imprinted on foil and survive the time in which it was produced. It is the voice of Edison reciting a nursery rhyme ("Mary had a little lamb ...").

Since then history has had a new standard, it has moved from the era of the *produced voice* to that of the *re-produced voice*. For the first time in 1877 it was possible to hear sounds, noises and voices produced not at the moment when they were perceived by the listener but at an earlier time, then recorded, stored and, on request, played. It was a revolution that radically changed the course of events and changed the way people were speaking. Until then speech was being modeled on what was heard. How many different voices will a speaker hear in his life in the era of the *produced voice*? If we consider that the speaker should be no more than a few meters from the listener, it is clear that the models of speech that everyone aimed to imitate, in order to achieve effective communication, were very few and very localized in terms of geographical area. In the era of *re-produced voice*, when voices arise continuously from so many objects (radio, television, movies, telephone, Hi-Fi, computers and so on) the number of voices that each of us hears every day is much higher than in the past. The models that came from different places, from different times, from different languages and cultures, thereby led to changes in the spoken languages as never before had happened. Thus, to give voice to objects has definitively been a great revolution, a revolution that produced, in the last century, changes that had not even occurred over a millennium.

Yet the history of attempts that more or less famous people have done to give voice to statues, idols or bronze heads is long and, for various reasons, not well known. The idea of the creation of a device capable of producing artificially the human voice has always had an adverse fortune. Whoever ventured on such a path was viewed with suspicion. Until the 17th century they were accused of practicing sorcery, and therefore persecuted and condemned; from the 18th century onwards they were often considered charlatans, illusionists.

A more fortunate fate has befallen those who have managed to design and manufacture their devices in other fields, however extraordinary they might appear to contemporaries. Just think of the burning mirrors arranged by Archimedes on the walls of Syracuse to defend the city from the ships of Marcellus, or to Archita of Taranto that in the 4th century BC builds a mechanical dove able to fly. No less amazing must have been the Greek fire, developed in the 7th century by Callinicus of Heliopolis, a device capable of launching firebombs. Neither Archimedes nor Archita nor Callinicus were later accused of sorcery.

Even inventions in the field of acoustics have been treated differently depending on whether the artificially produced sound was a musical composition or resembled the human voice. When Gerbert of Aurillac places in the cathedral of Reims an organ of his own invention which, thanks to the passage of steam from boiling water through metal pipes of different lengths, can produce wonderful sounds, contemporaries admire him. Conversely, when Gerbert builds a bronze statue that can emit articulate sounds, contemporaries fear him.

To understand the reasons for such a deep distrust in those who try to artificially produce voice, we must consider that voice is the only element that distinguishes man from other animals. Attitudes, instincts, even feelings that you may experience in humans can be identified in the behavior of other species. Only articulated language is exclusive to the human species. So to give voice to an object is far more amazing than to make it fly or emit melodious sounds. To give voice means to give life to something, to give a soul to the matter, and this is only granted to a god, but if it is a person who tries to do so, then it is surely the devil that secretly guides his hand.

For these reasons, to sketch the history of attempts of giving voice to objects, most often anthropomorphic, means scrolling the list of people who have been accused of practicing sorcery. For this reason it is a hidden history, and here I will try to trace it, focusing on some steps which seem particularly interesting and significant.

2. The two paths: the trick and the research

The history of attempts to give voice to objects developed along two well-defined paths, the path of the trick and the path of scientific research. The path of the trick is that in which the voice seems to spring from an object, that is a statue, a head or something of the sort, but in reality it is a simple transport of voice: the voice is produced by a hidden speaker and it is carried, by means of an artifice, in the direction of the talking head. Along the path of research, in contrast, the inventor tries to build a mechanism which, by imitating the phonatory apparatus, is able to produce an acoustic signal similar to voice. Between the two types of products there is an obvious difference: in the first case the machine speaks fluently and without difficulties, in the second case the result is very rough, the machine produces only some single sounds, maybe a short sequence of sounds. The devices of the first type were at most a source of curiosity in the fairs and theaters of illusionism, while those of the second type were studied in scientific laboratories. But which of the two routes led to modern talking machines? Contrary to what we might think, the answer to this question is not obvious.

3. The path of the trick

The idea of building a talking statue by the technique of voice transport goes back to very old age, to the period when the Aegean sea was full of oracles predicting the future, like the head of Orpheus in the Lesbos island. Or even earlier, at the time of the events narrated in the Old Testament. In Mesopotamia, the Chaldean priests were accustomed to question their idols, the *teraphim*, to predict the future. When questioned, the *teraphim* answered. The scene is very well represented in the painting "Consulting the Oracle" by John William Waterhouse [1]: we are inside a house, or a temple, the candles are burning, people are kneeling in front of the embalmed head that suddenly begins to talk. The trick will be explained many years later, in the 3rd century, by Hyppolitus in his Philosophumena [2]: the idol was made with wax and had the semblance of a skull. The skull was secretly connected with the windpipe of a crane or some other long-necked bird, so that the head seemed to speak. The accomplice spoke what the magician wished. When he wanted it to vanish, he offered incense, put a lot of coals around the head, the wax melted and the skull became invisible.

We must not smile at the credulity of the naïve spectators in front of the speaking head because, as we shall see, things have not changed over the centuries. Even today, although in a completely different way, the transport of voice continues to be practiced successfully. On the other hand, it continued to be used over the centuries in different places and times, even by famous people such as archbishops, philosophers and scientists. Starting from the 13th century, in fact, it seems that a large number of people possessed an oracle, most of the times made of bronze. There are numerous and contradictory testimonies relating to this period. A good reference is the work by Gabriel Naudé "Apologie pour tous les grands personnages qui ont été faussement soupçonnés de magie" [3]. The specific purpose of the work was to rehabilitate whoever, at various times, had been accused of sorcery. To save their reputations Naudé was forced to make a long series of extremely useful arguments, though today they would be considered inappropriate for being clearly groundlessness. Naudé enumerates all the testimonies that he can find: for him, who was Richelieu's and Mazarin's librarian, it must not have been difficult to find news, rumors, legends and medieval chronicles. Among these, many involve talking statues, such as those built by Robert Grosseteste, Roger Bacon, Albertus Magnus and Enrique de Villena. Here we focus on the statue that, in our view, is the most representative and whose story is in many ways most fascinating.

4. The Android of Albertus Magnus

There are numerous testimonies about the talking statue built by Albertus Magnus. Besides Naudé, Tostato Alphonso de Madrigalejo, who was bishop of Avila and lived in the 15th century also mentioned it [4]. So did the abbot of Velly in his "Histoire de France" [5], Joseph François Michaud in his "Biographie Universelle" [6] and many more. The idea of building a talking statue comes to Albertus during his years in Paris, where he gives courses at the College of St. Jacques. To listen to his lectures students come from all parts of Europe. The numerous monasteries scattered around the continent send their young students to follow the lectures by Albertus on theology, science, physics and philosophy. The crowd of students is so large that it cannot be easily accomodated. Thus, the lectures are held in a square that is not far from the College of St. Jacques in Paris: this is Place Maubert, near the Sorbonne, which owes its name to the contraction of *Maître Albert*. The students who throng to listen are mathematicians, astronomers, alchemists. Among them is Roger Bacon, the Franciscan monk and a scholar of physics and experimental sciences.

After his stay in Paris, Albertus goes to Cologne, and Bacon goes to Oxford. Both build a statue that is able to speak. That of Albertus, known as the Android, is placed in the cell of the Dominican friar and emits sounds like a human voice: the statue speaks. When Thomas Aquinas, a disciple of Albertus, arrives in Cologne, he is taken by Albertus to the cell. Thomas is terrified because he hears the statue speak very clearly. Believing it is the devil's work, he destroys it with a stick. Albertus comes in and says "in a minute you destroyed a work that cost me thirty years of work" [7].

To have an explanation of how the Android worked we must wait for the mid-17th century, when Athanasius Kircher explains how to build a statue that is able to speak [8]. Kircher is familiar with the anatomy of the ear and gives great importance to the shape of the cochlear canal, where, in his opinion, the sound is channeled and amplified. More than on the resonance phenomenon, his attention is focused on the reflection of the acoustic wave. Having established that the angle of incidence is always equal to that of the reflected wave, he concludes that it is possible to concentrate the sound at a given point through pipes with a particular shape: the *tubus ellipticus*, the *tubus conicus* and the *tubus cochleatus* (Figure 1).

The idea comes to mind to him while he is in a chemical laboratory equipped with a small canal for letting the smoke go out. Here he realizes that the rumors can go out of the locked room in an apparently miraculous way, following the same path of the smoke. Figure 2 shows his project to ensure that, through the pipes DES, the voices of those who speak in the lower chamber arrive in the chamber F.

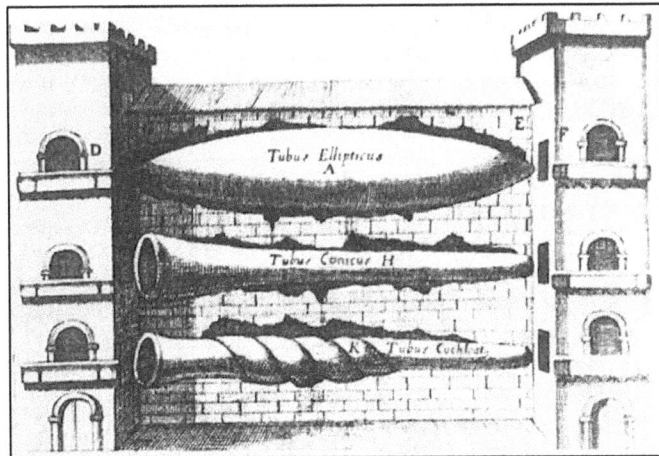

Figure 1: Tubus ellipticus, tubus conicus and tubus cochleatus (A. Kircher, Phonurgia Nova, Campidonae,1673, 100).

Figure 2: Tubi conici (A. Kircher, op. cit.).

But back to Albertus' talking statue and Kircher's explanations of it. The mechanism is shown in the Iconismus XVII, a splendid example of graphics, of imagination, of setting (Figure 3), and is explained by the Jesuit priest as follows: "[...] in the room ABCD, in which a tube-shaped concave [...] will be introduced in E, or in a vertical tube S, there is a statue which breathes from the mouth and the eyes [...]. This statue is placed in a precise place [...] so that the terminal part of the tubus cochleatus corresponds perfectly to the concave of the mouth; and you will have a statue that pronounces any word, perfect and complete. In fact, this statue will chat constantly, now uttering words of man, now of animal, now laughs [...] now you will hear singing, now crying, now shouting [...] with very strong general astonishment. Because in fact the orifice of the cochlear tube corresponds to a public place, all human words uttered outside, collected within the tubus cocleatus, come into the mouth of the statue. If dogs bark, even the statue will bark; if someone sings, the statue will respond in song. [...]. If we have the tubus cochleatus vertically, the thing will produce a better effect" [9].

Figure 3: Iconismus XVII (A. Kircher, Musurgiae Universalis, Roma, 1650).

According to Kircher's description, the history of Albertus Magnus' Android must have gone as follows.

Albertus built a hidden channel that connected his cell with another room. The extremity of the channel was formed by a metal head, probably of copper, in such a way that the sound appeared to come from the statue. The fame that accompanied and will accompany for centuries this ingenious gimmick shows that Albertus used it very often and with different people, probably to intimidate those with whom he was dealing, in order to establish a hierarchical relationship of dependency from the outset. When Thomas Aquinas arrives in Cologne, he is initiated by Albertus to the ritual and intentionally ("studiose" [10]) sent into the cell where there is the statue. As soon as Thomas goes in, the head starts to make sounds and to pronounce words. Evidently the channel connects the cell with another room, where Albertus starts talking to activate the mechanism. But alas, his pupil ("perterrefactus" [10]) breaks down the statue with a stick. Now the statue is silent because Albertus, who must have heard the noise from the other side of the cubicle, has run up and asked what is happening. Thomas responds candidly that it was he himself who crushed the statue because it was the devil's work. Albertus, controlling his anger, blurts out "Thomas, you have destroyed the work of thirty years!".

5. The end of magic

As we have seen, for many centuries the history of talking machines was a hidden history, since until the 17th century whoever tried to build a device emitting vocal sounds was accused of witchcraft and sorcery. The 17th century marks a total change of direction. If we want to refer to a specific time when that change occurs, we can say that the turning point was 1611, when the complete works of Giovan Battista Della Porta were published [11]. In Chapter I of Book XIX, in fact, the author wonders "whether a material statue can talk with some artifice". So, no more dark cells of monasteries (the most numerous to appear in the list of the accused were in fact ecclesiastics), but the possibility of creating a "talking statue" begins to be examined objectively, once back in the field of acoustics. Della Porta has no doubt about it: to those who accused Albertus Magnus to have made a talking statue by "the election of astrology", he answers firmly "But good God, how can a learned man believe this? How can the stars have strength to do these things? There are some who believe that he did for Magic Art. This I think less of all the things [...] but I think if he did, he did it for reasons of air" [12]. At this point Della Porta, trying to explain these "reasons of air", describes a mechanism that can capture a voice, store it in a tube and, on request, get it out and play it: it is the idea of a modern recorder, nearly three centuries earlier than Edison's phonograph. The voice finally becomes a material object that moves in space: "the words and the voice walk, so ordered by the air as they come out of the mouth". He imagines to build very long pipes of lead, a length of 200 or 300 steps, to pronounce some words at one end, then close the two ends of the tube and imprison the voice inside the tube. According to Della Porta, the voice will continue to bounce from one end to the other, so that "when you open the mouth of the pipe, voice comes out, as from the mouth of one who speaks".

Deprived of its magical atmosphere, the talking head with the trick of the voice transport will continue to be built and it will be an object of curiosity and fun in fair stands. Among the spectacles performed by illusionists, the most recurring is that of a head having the power of speaking and to answer questions asked by astonished spectators. The manners in which the experiments take place are different, but they can be grouped into two types of trick, one based on optics and one on acoustics. Among the latter, the one of the "Wooden Talking Head" was certainly original. A wooden head is suspended, by means of a chain made of brass, to the point of intersection of two arcs of metal wire (fig. 4).

The ends of the two wires are planted in the corners of a wooden box without lid and with the empty walls. In the mouth of the head there is a trumpet made of metal, and words and phrases can be clearly heard from it. To accentuate the impression that it is precisely the head that talks, the magician nears a lit match to the mouth of the trumpet and the flame fluctuates or even shuts down due to the breath emitted by the talking head. In reality, the trick is very simple.

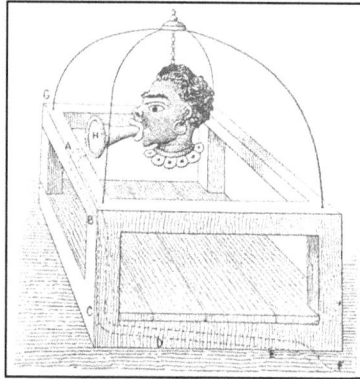

Figure 3: The wooden talking head (La Nature, n. 509, 1883, 221)

A hidden person speaks into a metal pipe of two centimeters in diameter. The pipe continues inside the wooden planks in the stretch FDCBA. At point A, the tube is curved in the direction of the center of the trumpet. The voice is therefore directed towards the bottom of the horn and, thanks to its conical shape, it is amplified and reflected back to the person who is located in front of the head.

Probably today the "Wooden Talking Head" would not impress anyone. However, if we consider that only a century ago it was one of the main attractions in theaters and fairs, we realize how, in past centuries, heads and bronze statues very similar to it were considered to be the devil's work.

6. The path of the research

Both Albertus Magnus and Roger Bacon drew much of their knowledge from the works of Arab philosophers, who in the field of natural science were far ahead of the Christian world, that was totally dedicated to prayer and asceticism, and where the only study to be allowed was that of theology.

Gerbert of Aurillac, who was elected Pope Sylvester II in 999, is the link between the Arab and the European cultures. For a number of favorable circumstances, but especially for its greed for knowledge, he uses what he has learned from his Arabian teachers in Seville and Cordoba in the field of acoustics, music and mechanics. This puts him in a very particular condition, as he is the first among Western Christians to become a disciple of the Arabs. While in the West the intervention of God or the devil is invoked to explain all that is incomprehensible, Gerbert approaches arithmetic and geometry, eagerly assimilating all that Arab culture has to offer. His life is a succession of incredible adventures that lead him to become the most powerful and feared man in the end of the first millennium. As with Albertus Magnus, there are numerous testimonies that Gerbert built a bronze head able to answer questions that are asked. Unlike what happens in the path of the trick, the statue of Gerbert can utter only two words: *etiam* for affirmative answers, and *non* for negative ones. It would be too long here to retrace in detail the history of this talking head. We can only say that the statue, which Gerbert used to impress and intimidate the naive audience, worked thanks to the strength of the steam of boiling water passing through two different resonators. The steam was originated by the presence of a *miliarius*, a vessel containing boiling water, on which device most of the mechanical experiments of the first half of the second millennium will be based. Gerbert had the idea of building a *miliarius* from the reading of a work that was well known in the Arab world, even if written by a scientist of Greek culture. The work, entitled *Spiritalia* had been written in the first century BC by an extraordinary man, who has been considered the first engineer: Hero of Alexandria. Among the many inventions described by Hero, one is how to give voice to a statue with the heat of the sun's rays. Hero will then be able to carry out his plan, giving rise to the most famous talking statue of the ancient world, known as the Colossus of Memnon [13].

In the 47th theorem of his *Pneumatica* Hero describes a trickling water fountain by the action of the sun's rays (figure 4). When the sun falls upon the globe EF, the air in it, being heated, drives out the liquid, which is then carried along the siphon G and passes through the funnel H into the pedestal ABCD. But when the globe is in the shade, having the air escaped through the globe, the tube sucks up the liquid again, and fills the void that has been produced. This takes place every time the sun falls upon the globe. In many theorems Hero explains how it is possible to get a sound by pouring water into a vessel. In general, if the base is connected with the outer air trough a very thin pipe, the inner air will be pushed into it abruptly giving rise to a hissing sound. Depending to what is at the end of the pipe different sounds can be produced: a bird's chirp, a dragon's hiss or a trumpet's play (figure 5).

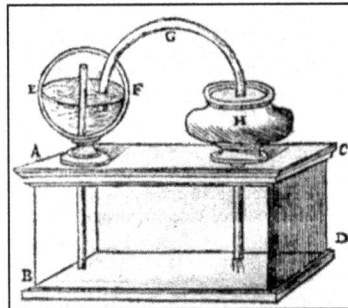

Figure 4: Hero's trickling water fountain by the action of the sun's rays. (Hero of Alexandria, Gli artifitiosi et curiosi moti spiritali di Herrone, 1589).

Figure 5: Statue playing the trumpet through the rays of the sun (Hero of Alexandria, op. cit.).

So the problem of how to create an airflow necessary for the production of a sound, a problem whose solution requires a complex mechanism of the lungs during phonation, is solved by Hero in a brilliant way.

To go back to Gerbert's statue, it had to have a very similar mechanism, so that the air was pushed inside. Instead of a single pipe, as in Hero's device, there should have been two. In fact the head answered affirmatively (*etiam*) or negatively (*non*). This means that it was able to produce two different sounds, which presupposes two resonators of different shape and size. From an articulatory point of view, which is the difference between *etiam* and *non*? The main difference lies in the fact that the first one has unrounded vowels, the second has a rounded one. Acoustically, this difference has enormous significance, since lip rounding involves the lowering of all formants, because a resonator with a hole of reduced output generates lower frequencies than one with a wider hole. Furthermore nasality causes a damping of intensity. These considerations lead us to believe that the head of Gerbert had a large cavity with a narrow outlet hole, suitable for producing a sound similar to [o], and one that was narrower but with a progressively larger output, useful to generate an [ɛ] sound. A system of shutters, operated by levers or by ropes, allowed the air to pass through one or the other resonator. Taking into account the different conformation of the two channels, it is likely that the first one was made inside the mouth and the other in one ear.

Probably the result was not great, but it should be considered that the listener was directed to choose between two possible answers (a methodology known today as the "binary forced choice task"). If we consider that all of this takes place a few years before 1000 AD, in a world filled with fear for all that appears inexplicable, we can imagine the effect on Gerbert's contemporaries.

Where might Gerbert's talking statue be today? Unfortunately we do not have an answer to this question, but, for those that are interested in seeing a very similar head. we suggest, to make a visit to St. Eulalia Cathedral in Barcelona. There it is possible to admire the head of the Moor (the "organ Carassa"), which testifies to the Christians' practice to take derision of the enemy by hanging a head depicting a Moorish king on the organ pipes. The head can now be operated by means of a rope: it opens its mouth and rolls its eyes. Unfortunately, not being connected to the pipe of a hydraulic organ, it is not able to emit any sound. However, for sure, it is truly spectacular (figure 6).

Figure 6: The head of the Moor, Barcelona.

7. The Abbot Mical

There is little information about the life of Abbot Mical. We know that he was born around 1730 and after finishing his studies and receiving holy orders, he had a benefit that allowed him to live quietly and modestly. He devoted all his time to the study of mechanics, the science for which he had a particular aptitude. Initially he built two robots that played the flute, later many others who played several instruments in order to form a full orchestra. This work, according to his contemporaries, "was able, for the masses, for the beauty of carved figures and for the perfection of the extremely varied, to beautify the largest hall." According to Louis Bachaumont, editor of the "Memoires Secrets" [15], the Abbot destroyed his work because he had been accused of having depicted naked figures. After this first attempt to produce musical automata, he built a bronze head able to articulate short sentences. Unfortunately, this time too the Abbot's modesty caused the talking head to have a fortune similar to that of the players' automata. In fact, the abbot showed it to an acquaintance who, betraying his confidence, wrote a letter to the "Journal de Paris" in which he praised the amazing invention. The abbot then destroyed the talking head as he considered it still too rough and imperfect.

After this episode, Mical built two new talking heads, the first real speaking machine. The Abbot presented his work at the Academy of Sciences on July 2, 1783 and the members of the jury recognized the great importance of his invention. In fact, as we can read on the engraving depicting the machine, "L'Académie des Sciences a dit dans son rapport que ces têtes parlantes peuvent jeter le plus grand jour sur le mécanisme de l'organe vocal et sur le mistère de la parole. La docte assemblée avait déclaré que cet ouvrage était digne de son approbation autant par sa nouvauté que par son importance que par son execution". The machine consisted of a sort of dome supported by four columns in Corinthian style, all decorated in Louis XVI style. In the middle of the canopy there were two heads placed on a small gallery also supported by Corinthian columns. In front of these small pillars there was a cloth with the words spoken by the two robots (Fig. 7).

The head on the left says: "Le roi donne la paix à l'Europe". At this point the second head replies "La paix couronne le roi de gloire". The first head then says "Et la paix fait le Bonheur des peuples. Oh roi adorable/père de vos peuples/Leur Bonheur fait voir à l'Europe/La gloire de votre trône". According to the testimonies of those who had the chance to listen to it, Abbot Mical's machine was stunning, and its mechanism was so explained: "Les têtes recouvraient une boîte creuse, dont les différentes parties étaient rattachées par des charnières, et dans l'intérieur de l'auteur laquelle avait disposé des glottes artificielles, de différentes formes, sur des membranes tendues. The air, passant par ces glottes, allait frapper les membranes, here rendaient des moyens sons ou aigus, et de leur espèce d'une combinaison résultait imitation très-imparfaite de la voix humaine". It should be noted that, wisely, Abbot Mical had highlighted on the drape the text of the sentences that the listener should recognize, thus helping them in the identification of the text.

Unfortunately for the Abbot, a police lieutenant, Jean-Charles Lenoir wrote a negative report, because he feared that it was a trick, a deception. The authorities in Paris, following his advice, decided not to buy the talking machine. The Abbot was deeply disappointed, he abandoned his research and retired to private life. He died a few years later, in 1789, and we have lost track of his machine. Someone says that the Abbot, disappointed, destroyed it voluntarily. Is it true? In our view, things could have gone differently and maybe it is still possible to find the machine. An interesting clue is given by the testimony of Jean Etienne Montucla, the author of an important "Histoire des mathématiques", a friend of D'Alembert and Diderot, and a regular attender at the Encyclopedists' meetings.

Figure 7: The talking heads of Abbot Mical (La Nature, 1905, n.1667).

According to Montucla Mical's machine was by no means destroyed: the Abbot sold it for a considerable sum, it seems, to a foreign nobleman [16]. Reading carefully the testimonies relating to Mical's talking heads, indeed a "foreigner" seems to have certainly had the opportunity to see them, listen to them and appreciate them. The event took place June 18, 1783, when the abbot decided to invite to dinner in his home, in *rue du Temple*, two members of the Academy of Sciences that soon would have to judge his work. One of these, Barthélemy Faujas de Saint Fond, on the morning of June 18, writes a note to another member, an American, to remind him of the Abbot's invitation. He fears that the eminent American diplomat, on a visit to France to report the events of the American Revolution, may have forgotten -because of his many commitments- the unknown Abbot's dinner invitation. So that night the two men admired the work of the abbot and were fascinated by it. That American was Benjamin Franklin.

Where can Abbot Mical's talking heads be today? Unfortunately, this question does not have an answer yet.

8. Conclusions

As we have seen, the history of talking machines followed two different paths, that of the trick and that of the research. In the first case, the voice was actually produced by a hidden subject and transported through an artifice to the place where there was a head or a statue. In the second case, the inventor tried to build a mechanism that in some way was able to imitate the complex mechanisms of phonatory organs during the production of speech. Which of these two paths led to the synthesized voice of modern talking machines? Contrary to what one might think, it is not only the course of research that has led the most progress in this field, but also that of the trick. The synthesis for diphones, in fact, is based on a database of speech segments actually produced by a speaking subject, suitably cut, stored and reproduced on demand in the appropriate sequences. The results are every day more and more satisfying and amazing, certainly not comparable to the rough and naive attempts of their ancient predecessors, but the story of talking machines is definitely history that is far from being over.

References

[1]   Waterhouse, J. W.: Consulting the Oracle, Tate Gallery, London, 1884.

[2]   Hyppolitus of Rome: Philosophumena or the Refutation of All Heresies, book IV, The Diviners and Magicians, around 220 A.D.,103.

[3]   Naudé, G.: Apologie pour tous les grands personnages qui ont été faussement soupçonnés de magie, Paris, 1625.

[4]   Tostato, A.: Commentaria in secundam partem Numerorum, Venise, 1615.

[5]   Velly, Abbé de: Histoire de France, Paris, 1761.

[6]   Michaud, J. F.: Biographie Universelle, Paris, 1854.

[7]   "Opus triginta annorum destruxisti". In: Majoli Simonis Episcopi Vulturariensis, Dierum Canicularium tomi septem. Colloquis quadraginta, Francofurti, 1642. Author's translation.

[8]   Kircher, A.: Phonurgia Nova, Campidonae,1673.

[9]   Kircher, A.: Musurgiae Universalis, Romae, 1650, 306. Author's translation.

[10]   Majoli Simonis Episcopi Vulturariensis, Dierum Canicularium tomi septem. Colloquis quadraginta, Francofurti, 1642, 315.

[11]   Della Porta, G. B.: Della Magia Naturale, Napoli, 1611.

[12]   Della Porta, G. B., op. cit, 688. Author's translation.

[13]   Pettorino, M.: "Memnon, the vocal statue", in Proceedings of the 14th International Congress of Phonetic Sciences (ICPhS), San Francisco, 1999.

[14]   Hero of Alexandria: Gli artifitiosi et curiosi moti spiritali di Herrone, 1589.

[15]   Bachaumont, L.: Mémoires Sécrets ou Journal d'un observateur, Paris, 1884.

[16]   Reported in Michaud, J. F. op. cit., 186.

Kempelen vs. Kratzenstein –
Researchers on speech synthesis in times of change

F. Brackhane

Institut für Deutsche Sprache (Mannheim, Germany)
brackhane@ids-mannheim.de

Abstract: One was a distinguished natural scientist and engineer, the other a self-taught scientist and vilified as a conman: Christian Gottlieb Kratzenstein (1723–1795) and Wolfgang von Kempelen (1734–1804). Some of the former's postulations on human physiology and articulation of speech proved wrong in later years. Most of the latter's theories are considered applicable even today. The perhaps most contrasting approaches to speech synthesis during the 18th century are linked to their names. There are many essential differences between their approaches which show that these two researchers were not only representatives of different schools of thought, but also representatives of two different scientific eras. A speculative and philosophical approach on the one hand versus an empirical and logical approach on the other hand. Both Kratzenstein and Kempelen published books on their research. But while the "Tentamen" [4] of the physician Kratzenstein remains rather vague and imprecise in its descriptions of vowel production and synthesis, the "Mechanismus" [8] of the engineer Kempelen shows much more precision and correctness in almost every respect of human speech and language. The goal of this paper is to discuss the differences between these two contemporaneous researchers on speech synthesis and to compare their theories with present-days findings.

1. Speech research in the 18th century

During the lifetime of Christian Gottlieb Kratzenstein and Wolfgang von Kempelen a profound change in science in general and speech science in particular took place. Although the revolutionary ideas of the age of enlightenment were not new the traditional way of science still had a lot of supporters. For instance, the debate about the question whether speech was given by god or invented by human lasted until the 19th century. Also there was no conclusive evidence which organs were responsible for speech production in which way. In 1667, the philosopher and polymath Franciscus Mercurius van Helmont (1614–1699) published his book "Alphabeti vere naturalis". In this he postulated that speech was given by god and there was a proto-language. He argued for Hebrew being this language because of the "anatomical" shape of the Hebrew letters [1]. While theories such as these may seem strange to us today, they were still very common in the late 18th century. While Kratzenstein was a proponent of these traditional ways of thinking, Kempelen was a representative of the "modern" way of critical and empirical science. Nonetheless both their approaches to speech synthesis were very similar in some ways.

2. Christian Gottlieb Kratzenstein (1723-1795)

2.1. Biography

It is unknown when Kratzenstein was born, but he was baptized on February 2, 1723 in the city of Wernigerode (Germany) where his father Thomas Andreas Kratzenstein was a jurist and mayor. From 1742 on he studied humanities as well as natural sciences with a focus on medical science in Halle and graduated in 1746. Subsequently he taught as private lecturer in Halle until he was placed at the academy of sciences in St. Petersburg (Russia) by Leonhard Euler in 1748. In 1753 he moved to Copenhagen (Sweden) and stayed there until his death on July 7, 1795. Just one month earlier a major fire had destroyed large areas of Copenhagen including Kratzenstein's home with nearly all his letters, books and instruments. [2]

2.2. The scholar

While he had taught in Halle various aspects of medicine and natural science in general, since 1748 Kratzenstein officiated in St. Petersburg as a professor of mathematics and mechanics. Since 1753 he was professor for experimental physics in Copenhagen, a position which was set up especially for him. During those years, he published around 50 publications that show his very broad and versatile interests.

Although he was a natural scientist by education and an appointed professor, Kratzenstein was very well known in his lifetime for improving measuring instruments by others and inventing a few new ones. Because of this ability he was asked for advice by a number of other scientists all over Europe and esteemed for his expertise. But all his devices were not end in itself but served as support of his research on different topics. None of his instruments were intended as gadgetry but often as appliances for making findings of his research comprehensible to laymen.

2.3. The vowel organ

The academy of sciences at St. Petersburg organized an annual scientific competition. For the competition of 1780 the characteristics of the five vowels A E I O U had to be explained. Additionally a suggestion for a synthesis of these vowels based on organ reed pipes had to be given [4: title page]. The verbalization of this topic originates from Leonhard Euler who had broached this issue in a collection of problems for the Petersburg academy in 1765 and also in his "Briefe an eine deutsche Prinzessinn [sic!]" [5: 236; 6: 235[1]].

Kratzenstein submitted his paper anonymously under the codeword "plus ultra". Apparently his submission was not the only one because it was named "No. 2" by the jury [3: 111]. But there is nothing known about any other submitter.

In the first part of his text, Kratzenstein developed a theory of the human voice in general and of vowel production in particular. Obviously he had worked on these problems for some 10 years prior [7: 157]. Today this theory seems to be very dubious because it was widely based on a philosophical approach without any anatomical observations. But in fact, the physically based approach to determine human speech parallel to the sounds of animals or musical instruments was a new perception.

Kratzenstein refused the analogy of the voice to a string instrument (as represented by Antoine Ferrein (1693-1769) for instance) and tended towards the opinion of Denis Dodard

---

[1] [6] was first published in a French version from 1668 onwards.

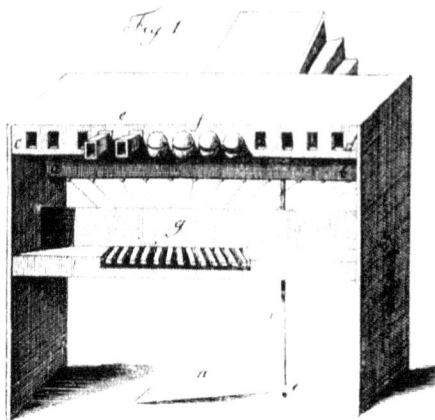

Figure 1: The second prototype of Kempelen's speaking machine
which seems to have great similarity to the vowel organ of Kratzenstein [9: Tab. XVII].

(1634-1707) who had described the human voice as a wind instrument. Therefore he also denied Ferrein's theory that it is the vibration of the vocal cords which is responsible for the pitch of the voice [7: 157]. Instead of that, Kratzenstein named the epiglottis as the crucial part of the human anatomy for the production of the voice. In the second part of the treatise a description of an organ based vowel synthesizer was given. This description, however, is very vague so that it is not possible to get any precise idea of the design of his vowel organ.

Along with his paper, Kratzenstein presented a little organ with which he wanted to synthesize the five vowels named in the competition. Sadly this instrument seems to be lost and no drawing or other description is known.

According to the transcripts of the academy, there was a second vowel organ. Apart from Kratzenstein, an instrument maker named Kirsnick had made that already in 1779. But this was not as perfect as Kratzenstein's was. Both instruments seem to have been very similar, containing keyboards and bellows operated by the feet of the player [3: 111]. Neither the organs nor any drawings of them have survived. But a drawing of a very similar prototype of a vowel synthesizer based on organ pipes is given in Wolfgang von Kempelen's "Mechanismus" (Fig. 1).

The works of Kratzenstein and Kirsnick seem to be very closely related to each other. While the transcripts of the academy date Kirsnick's invention first, the very popular organist and inventor Abbé Vogler, who later cooperated with Kirsnick, claimed that Kirsnick's vowel organ was based on Kratzenstein's drawings. According to Vogler, Kirsnick was the second candidate in 1780 and gained the second prize for his instrument [8: 128].

What did the vowel organ look like? It consisted of a small wind chest with five pipes operated by a little keyboard and foot pumped bellows. The pipes were reed pipes (except the one for "I" being a flue pipe like a recorder). While traditional reed pipes of an organ show great similarity to the functionality of a clarinet, Kratzenstein used a mutation unknown in Europe unto that time: free reeds. Those reeds have been used for many centuries in Asia in the musical instrument named sheng, but it is not certain whether Kratzenstein knew such instruments. In contrast to a regular reed pipe where the reed rests on its shallot and can only swing upwards, in a free reed pipe the reed is only attached to a frame and can therefore swing upwards and downwards. Because of this, the produced sound of a free reed pipe consists of sinusoidal waves instead of sawtooth waves.

For each of the five pipes Kratzenstein designed an individually shaped bell which modulated the sound in such a way that one of the five vowels could be heard (Fig. 2). Instead of brass as

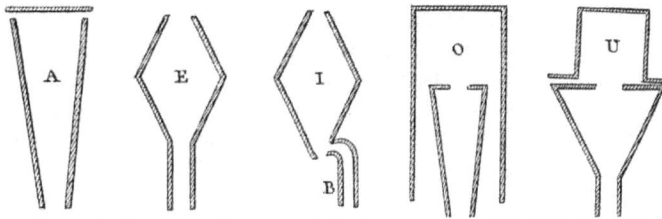

Figure 2: Shapes of Kratzenstein's five pipe bells [12: Tab. XXVI].

usual for reeds in organ pipes Kratzenstein used ivory or baleen hoping this organic material would support the humanlike sound of the pipes [3: 113].

This organ supposedly could pronounce the five vowels very clearly. In addition, it could produce some other speech sounds [3: 111].

3. Wolfgang von Kempelen (1734-1804)

3. 1. Biography

There are many legends and half-truths surrounding Kempelen and his work. Most of them originate in the early 19th century when biographical articles about him (and many others) were mostly based on hearsay. Some of those imprecise or wrong facts were adopted by later literature so that Kempelen's biography written by Reininger (2007) is the first really reliable resource, which has regrettably not yet enforced [10].

Wolfgang von Kempelen was born on January 23, 1734 in Preßburg (then Hungary, now Bratislava in Slovakia). His father Engelbert (von) Kemp(e)len served as a civil servant for the Austrian administration in Hungary for which he was ennobled in 1722. Since 1755 Wolfgang von Kempelen himself served as a civil servant. First he worked as a project manager at the "Hofkammer" (finance office) at Preßburg. In the following years he went through several positions in the Hungarian administration but he acted also as managing director of various state-owned manufactories in Hungary. In 1776 he was responsible for the relocation of the University of Trnava (now Slovakia) to Budapest (Hungary). In 1785 after a biennial leave during which he travelled through Germany, France and England Kempelen became an ethnarchy council member in Budapest and two years later "Hofrat" for the region of Transylvania at the administration in Vienna until his retirement in 1798. Kempelen died on March 26, 1804.

3.2. The inventor

During his entire life Kempelen was engaged in the development or improvement of technical appliances which were however only connected partly with his duties. He developed and supervised, for example, new pumping stations for the castles in Schönbrunn and Budapest. In 1774 Kempelen designed a special bed for the ailing empress Maria Theresia, whose personal goodwill had attended him throughout. Since 1777, he worked on a steam-engine prototype which he improved until 1793. James Watt knew about Kempelens work and thought it was very good. In 1779, he developed a sort of typewriter for blind people, especially for the blind singer Maria Theresia von Paradis, whose teacher he was. In addition, he wrote some plays and made copperplate engravings.

Kempelen had become famous beyond Austria overnight through the construction of his mechanical "chess player" ("chess Turk") which he had constructed in 1769, allegedly within half a year. This invention has been misunderstood in many ways by the contemporaries.

Many of them believed that this device was an android but Kempelen himself never claimed this (he just did not say anything about the question how it worked generally). For the following 20 years, journalists and publicists tried to solve the mystery of this invention splitting up into two groups: Those who wanted to prove objectively and logically that the "chess player" had to be a mechanical device operated by a human and those who spoke out against Kempelen as a cheater in general.

3.3. The speaking machine and the "Mechanismus der menschlichen Sprache"

In 1783 when Kempelen started his journey through Europe to exhibit his "chess player" especially in Leipzig, Paris and London his fame had suffered a lot. This was due not least to the book of his friend Karl Gottlieb von Windisch (1725-1793) about the "chess player" which was supposed to be an advertisement in fact and therefore published in German, English, French and Dutch. But instead of advertising the "chess turk", this publication confirmed Kempelen's critics so that the journey did not become a great success and Kempelen stood as an impostor in the end. Because of this, he decided to publish a book himself about his second great invention, the speaking machine which he had engineered parallel to the "chess player" and also presented during his journey. Unlike the latter, the speaking machine was not an (although excellent) prestidigitation but a deeply serious outcome of natural scientific observations. But because knowledge about human anatomy and speech production was not widespread to this time, Kempelen had to proof his invention basically. So primarily the "Mechanismus" is an explanation of the speaking machine which is given in chapter 5. Chapters 1–4 give the necessary backgrounds for understanding the structure of the speaking machine. They deal with questions of how one can define speech, of its origins (developed by human or given by god), of the physiological requirements for producing speech and of the speech sounds. Although this book is in fact a trial of justification for the construction of a mechanical showpiece of an amateur it nonetheless became a very notable scientific paper on speech. Kempelen was a striking progressive scientist and therefore a characteristical representative of the age of enlightenment. The very most of his theories on speech production were accepted as true to date.

The speaking machine was a continuation of his typewriter for blind people in a way. It was Kempelen's idea to provide so called deaf-mutes with something like prosthetic devices for communication. Because of this, he was not satisfied with only some vowels or other speech sounds. He wanted to create a mechanism that could synthesize every speech sound [9: 389]. Kempelen recognized as one of the first the essential importance of coarticulation for speech production and synthesis [9: 197, 407]. So, unlike Kratzenstein, he designed his speaking machine very closely to the human anatomy and tried to imitate the movements of the mouth as possible.

The speaking machine consists of three major parts which are fixed on a small wooden windchest (Fig. 3): Bellows as representation of the human lungs, a reed pipe derived from a pipe organ as source of sound (according to Kratzenstein the reed pipe was furnished with a reed made of ivory) and a rubber funnel for the mouth. This "mouth" contained no analogies for the human tongue and teeth. As a substitution Kempelen installed two separate devices which are independent from the "mouth": A modified mouthpiece of a recorder for producing a SH-like sound (Fig. 3 bottom) and a little tin for producing a S-like sound (Fig. 3 top). Both are operated by levers on top of the machine. The machine is operated by both hands and the right arm: While the right elbow pushes the bellows, the palm of the left hand covers the rubber funnel more or less to modulate the sound. The right hand is needed for controlling the nose cavity and the levers. Production of voiceless sounds is possible only very restricted because it is not possible to abduct the reed of the reed pipe.

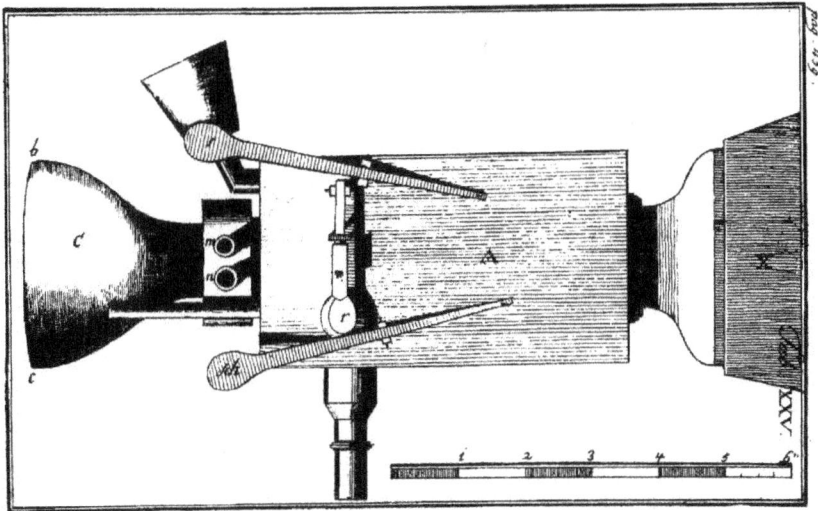

Figure 3: Top view of Kempelen's speaking machine [9: Tab. XXV].
From right to the left: Bellows (X), windchest (A), nose cavity (m, n), rubber funnel (C).

4. Discussion

Both Christian Gottlieb Kratzenstein and Wolfgang von Kempelen had remarkable abilities for constructing mechanical devices although they were not skilled mechanics. Both were representatives of the Enlightenment, although their perceptions differed notably. Although both probably were able to get aid by professional precision mechanics at least the conceptual design and perfection of their work was their own merit. Especially Kratzenstein was noted by other scholars for his broad knowledge in scientific as well as in technical matters. But also Kempelen must have been known for his very broad abilities as suggested by his very manifold functions in the administration of the Austrian monarchy.

Ostensible both constructed prototypes of speech synthesis devices based on technologies adopted from the pipe organ. Both seem to have failed in the end. On closer inspection, however, there are differences between both approaches which could hardly be greater. While Kratzenstein was a physician by profession, Kempelen was a manager and technician who had acquired his physiological knowledge self-educated. Nonetheless Kempelens empirically obtained insights of the human physiology and mechanisms of speech production were much closer to modern day's knowledge than the philosophically and theoretically based theses by Kratzenstein.

The particular construction of the two synthesis apparatuses reflect two fundamentally different approaches which result from very different theories of speech production. The theories of vowel articulation should stand exemplary for those great differences: Kratzenstein recognized the nature of speech sounds as similar to light; vowels were produced by multiple reflected "sound rays", generated by complex shapes of the mouth (Fig. 4 left). Kempelen however had a less abstract but more anatomically based view. He specified two independent articulatory movements to be essential: The opening of the mouth on the one hand and the opening of what he referred to as the "tongue channel" on the other (Fig. 4 right).

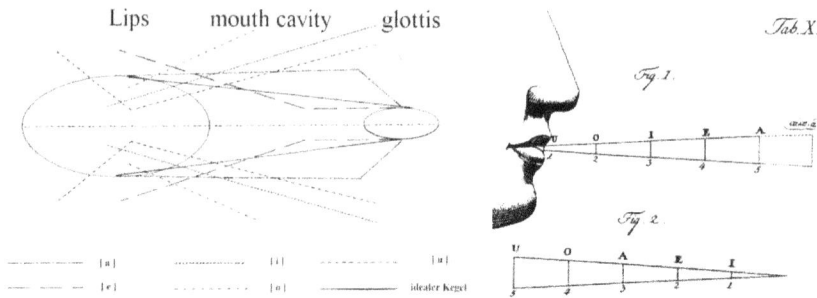

Figure 4: Kratzenstein's (left) and Kempelen's (right) theory of vowel articulation
[11: 572; 9: Tab. X]

Kratzenstein was very well known as an outstanding toolmaker and as improver of mechanical instruments and tools in particular ("refining by duplication" [2: 127]). According to his other scientific work, Kratzenstein tried to put the understanding and the explanation of speech production into abstract and geometrical forms. For him anatomical observations were just a starting point for the development of a mechanical device. Kempelen's theory of vowel production was completely different and much closer to our modern perception. He observed the anatomical conditions of the human vocal tract and wanted to simulate them as precisely as possible. Through this simulation he wanted to produce humanlike speech. Therefore his approach was one of indirect speech synthesis in contrast to Kratzenstein's.

It is not clear today why both Kratzenstein and Kempelen furnished their reed pipes with reeds made of ivory which is a completely uncommon material for reed pipes. It is conceivable that both hoped to generate a more "animated" sound by using an organic material.

Seen from today the approach of the "dark horse" Kempelen was much more applicable and promising than the one of the famous scientist and tool maker Kratzenstein. He had recognized the absurdity of the theological based theories by Helmont[2] and others very clearly and used a modern, empirical way of research instead. Nevertheless his prototype of a speech synthesis device shows great similarities to Kratzenstein's relating to its basic functional principle. Even though his attempt to make a fully functional speech synthesis device were not entirely successful, Kempelen established a new, critical approach to speech research for which he is sometimes called the founder of experimental phonetics. Kratzenstein, by contrast, is remembered as a scientist who tragically failed to develop modern technology by means of traditional ways of natural science.

Acknowledgements

Many thanks to Dr. Beata Trawiński and Ekaterina Most who translated particularly interesting sections of [3] into German. Many thanks also to Anette Klepp for improving the English version of this paper.

---

[2] "His [Helmont's] overheated fantasy foists curves and embellishments on the tongue that not only does it not take on in the letter concerned, but is not capable of taking on in any circumstance." [9: 144] (translation: Richard W. Sproat)

References

[1]   Helmont, F. M.: Alphabeti vere naturalis hebraici brevissima delineatio, quae simul methodum suppeditat, juxta quam, qui surdi nati sunt, sic informari possunt, ut non alios saltem loquentes intellegant, sed & ipsi ad sermonis usum perveniant. Sulzbaci: Lichtehthalerus. 1667.

[2]   Splinter, S.: Zwischen Nützlichkeit und Nachahmung. Eine Biografie des Gelehrten Christian Gottlieb Kratzenstein (1723-1795). Frankfurt (Main): Lang. 2007.

[3]   Koplevič, J. Ch.; Cverava, G. K.; Grant, K.: Christian Gottlieb Kratzenstein. 1723-1795. Leningrad. 1989.

[4]   Kratzenstein, Ch. G.: Tentamen resolvendi problema ab illustri Academia Imperiali scientiarum Petropolitana ad annum 1780 publice propositum. Petersburg. 1781

[5]   Juškevich, A. P.; Winter, E.: Die Berliner und die Petersburger Akademie der Wissenschaften im Briefwechsel Leonhard Eulers. Vol. 3. Berlin: Akademie-Verlag. 1976.

[6]   Euler, L.: Briefe an eine deutsche Prinzessinn über verschiedene Gegenstände aus der Physik und Philosophie. Aus dem Französischen übersetzt. Vol. 2. Leipzig: Junius. 1773.

[7]   Ohala, J.: Christian Gottlieb Kratzenstein: Pioneer in speech synthesis. Proc. 17th International Congress of Phonetic Sciences (Hong Kong), pp. 156-159. 2011.

[8]   Vogler, G. J.: Ueber Sprach- und Gesang-Automaten. Sammlung einiger in dem Frankfurter Museum vorgetragenen Arbeiten. Heft 1, S. 118-130. Frankfurt. 1810

[9]   Kempelen, W.: Wolfgangs von Kempelen k. k. wirklichen Hofraths Mechanismus der menschlichen Sprache nebst Beschreibung seiner sprechenden Maschine. Wien: Degen. 1791.

[10]   Reininger, A.: Wolfgang von Kempelen – Eine Biografie. Wien: Praesens. 2007.

[11]   Gessinger, J.: Auge & Ohr – Studien zur Erforschung der Sprache am Menschen. Berlin: De Gruyter. 1994.

[12]   Young, T.: A course of lectures on natural philosophy and the mechanical arts. Volume 2. London: Taylor and Walton.

[13]   Brackhane, F.: „Kann was natürlicher, als Vox humana, klingen?" – Ein Beitrag zur Geschichte der mechanischen Sprachsynthese. Saarbrücken: Dissertation. 2015.

„Eine Kempelensche Sprechmaschine".
New insights in speaking machines
in the late 18th and early 19th centuries

Silke Berdux

Deutsches Museum, Munich, Department of Musical Instruments
s.berdux@deutsches-museum.de

It was in 1791 that Wolfgang von Kempelen published the *Mechanismus der menschlichen Sprache nebst der Beschreibung seiner sprechenden Maschine* in Vienna, the last part of which is dedicated to the description of the speaking machine he had worked on for many years. In some parts of his book Kempelen refers to ideas he hadn't been able to put into practice at that time and asks the readers to further develop the machine.

It was already known that several machines were produced subsequently [2, 3, 6, 7]. For example, Johann Wolfgang von Goethe mentioned in June 1797 in a letter from Jena: "Kempelens Sprechmaschine, welche Hofrat Loder besitzt und die zwar nicht sehr beredt ist, doch aber verschiedne [sic] kindische Worte und Töne ganz artig hervorbringt, ist hier durch einen Tischer [sic] Schreiber, recht gut nachgemacht worden". In 1806 the well-known medal maker and sculptor Leonhard Posch presented a speaking machine in Berlin. In contemporary sources it is said that it had been built on the basis of Kempelen's machine and was very similar, but improved. Finally, the British physicist Charles Wheatstone optimized Kempelen's machine and built an apparatus which he presented in 1826.

Copying the well-known speaking machine in the Deutsches Museum in Munich, often called "the speaking machine by Wolfgang von Kempelen", was the reason for research about its history, context and attribution which included speaking machines of the late 18th and early 19th centuries [8]. They brought new insights in the speaking machine of Justus Christian Loder mentioned by Goethe, about which little was known, despite the prominence of both the writer and the owner – Loder being a famous doctor of the late 18th and early 19th centuries –, so that its story can be told now. They also brought insights on other unknown speaking machines built on the basis of the one described by Kempelen.

The recent findings show that the interest in speech synthesis, manifested since the mid-18th century in the ideas and works of Erasmus Darwin, Leonhard Euler, Christian Gottlieb Kratzenstein und Abbé Mical, became more common around 1800. There are several persons dealing with this topic – presumably stimulated by the publication of the *Mechanismus* –, constructing the machine described by Kempelen, improving it and using it for their purposes. Speaking machines can be found in very different contexts showing that the interest in speech synthesis expanded in diverse areas. These include anatomical and physical collections, mechanical experiments becoming part of a *Kunstkammer* and the partly spectacular presentations of public science.

References

[1]   Kempelen, Wolfgang von: Wolfgangs von Kempelen k.k. wirklichen Hofraths Mechanismus der menschlichen Sprache nebst der Beschreibung seiner sprechenden Maschine. Wien: Degen 1791.

[2]   Niemann, W.: Sprechende Figuren – Ein Beitrag zur Vorgeschichte des Phonographen. Geschichtsblätter für Technik und Industrie 7 (1922), 2-30.

[3]   Pompino-Marschall, Bernd: Wolfgang von Kempelen und seine Sprechmaschine. Eine biographische Notiz zum 200. Jahrestag der Publikation seines „Mechanismus der menschlichen Sprache". Forschungsberichte des Instituts für Phonetik und Sprachliche Kommunikation der Universität München 29 (1991), 181-252.

[4]   Gessinger, Joachim: Auge & Ohr. Studien zur Erforschung der Sprache des Menschen 1700-1850. Berlin: de Gruyter 1994.

[5]   Felderer, Brigitte (ed.): Phonorama. Eine Kulturgeschichte der Stimme als Medium. Katalog der Ausstellung im ZKM Zentrum für Kunst und Medientechnologie Karlsruhe, Museum für Neue Kunst, 18. September 2004 - 30. Januar 2005. Berlin: Matthes & Seitz 2004.

[6]   Brackhane, Fabian: Die Sprechmaschine Wolfgang von Kempelens – Von den Originalen bis zu den Nachbauten. Phonus 16 (2011), 49-148 (Forschungsberichte des Instituts für Phonetik der Universität des Saarlandes).

[7]   Brackhane, Fabian: „Kann was natürlicher, als Vox humana, klingen?" Ein Beitrag zur Geschichte der menschlichen Sprachsynthese. Saarbrücken 2015 (Phonus. Berichte zur Phonetik Universität Saarbrücken 18).

[8]   Berdux, Silke (ed.): Der Sprechapparat im Deutschen Museum in München. München: Deutsches Museum (forthcoming).

Kratzenstein's vowel resonators – reflections on a revival

Christian Korpiun

Most commonly, Kratzenstein's phonetic works are discovered by way of the widespread pattern drawings of his resonators in monographs and essays.

The following reproduction is taken from the eighth issue of the popular scientific book *A System of Natural Philosophy* of 1845, which appears to be the origin of the drawings and their later variations.

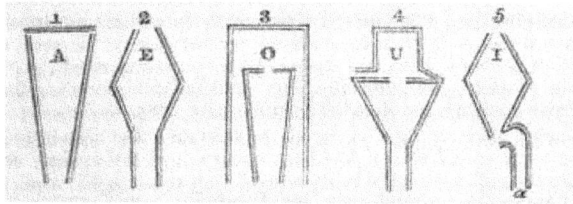

Pattern drawings of the Kratzenstein resonators [1]

They are usually accompanied by terse explanations of the following kind: 'In 1781, Christian Kratzenstein, a German scientist, was awarded a prize at the Saint Petersburg Academy of Arts for his presentation of five resonators that served to produce the five vowels a, e, i, o and u. Shortly afterwards, Wolfgang von Kempelen constructed a speech machine which was able to produce syllables.' This is often followed by the mention of Charles Wheatstone, who reportedly experimented with a reproduction of Kempelen's machine in the first half of the 19th century. These are, in short, the few facts that are commonly disseminated about Kratzenstein and his vowel resonators.

1. Starting practical

On the basis of these rather scant notes I began in 2005 to design resonators from sheet zinc from the above sketches and a comparison with the corresponding sections through the vowel tract. In treatments of the subject 'vowel production' in the course of several lectures on the physiology of language, the accompanying demonstration of the resonators served to

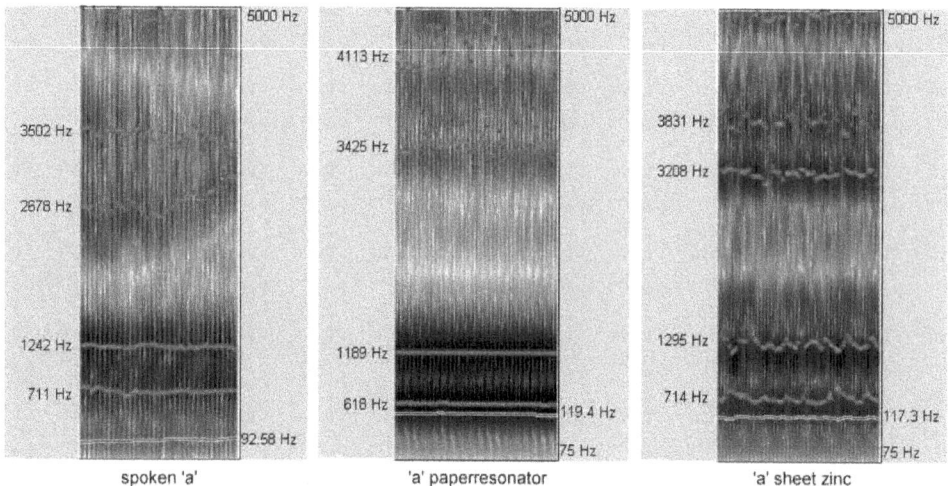

spoken 'a'   'a' paperresonator   'a' sheet zinc

considerably raise the students' interest in phonology. Subsequent measurements with the vowels stimulated by a saw tooth frequency resulted, over and above the subjective hearing impression, in a distribution of the formants that correlated surprisingly well with the spoken vowels. [2]

Set of resonators from sheet zinc

2. Using the 'Tentamen resolvendi problema ab Academia Scientiarum Imperiali Petropolitana ad annum 1780 publice propositum'

An original copy of the 'Tentamen' became available only in 2008 when Wilfrid Braun, a former colleague, retrieved its location in the collection of the Goettingen State and University Library. I then arranged its digitization – including the plates – in order to make the text available on the Internet.

A first reading of the 'Tentamen', especially of §§ 9. and 10. gives the impression that it was written to provide directions for the construction of resonators. In § 9. Kratzenstein assembles a list specifying the aperture dimensions of the vowel tract. These had been determined with the aid of a flat chip of wood which was put upright between the teeth or lips; for particular widths it could then be ascertained which vowels could be correctly articulated by observing the corresponding opening.[3] Kratzenstein points out that these measures are neither exact nor observable simultaneously.[4] In the same context we also find an early mention of the fact that the vowels are involuntarily pronounced with an initial consonant.

The data probably originate from previous works by Kratzenstein, in which he had occupied himself with phonetic as well as anatomical features of the vowels some time before the prize question.[5] In paragraphs 11 to 24 he deals with Denis Dodard and Antoine Ferrein as well as with the advancement of their theses by Albrecht von Haller [6] in order to develop a view of the production of the vowels from a physician's point of view.

Section 2 of the 'Tentamen' is headlined: 'De construendis fistulis, vocales a, e, i, o, u enunciantibus' (about the construction of pipe-whistles which express the vowels a, e, i, o, and u). To begin with, it must be stressed that Kratzenstein did not conceive of the construction of the vowel resonators as an imitation of the vowel tract. Instead, he developed the forms by extensive experiments. It is not surprising that these basically reproduce form principles of the vowel tract, as we will see in the case of 'e'. Their measures, however, are derived acoustically from the sizing of the 'e' to correspond to the tone C' (at that time 247 hertz). For the height of a cone Kratzenstein indicates a length of about 3 'thumbs'[7] (foot/12). Sections 26 to 30 describe in more or less detail the construction of the resonance bodies prepared for the vowels. [8] This is supplemented by the original drawings displayed in the appendix.

	Larynx	Lingua	Apertura viae palatin.	Apertura dentium.	Apertura labiorum.
A.	Latera ejus parum deprimuntur et dilatantur. Epiglottis parum elevatur.	Apex ad radices dentium maxillae inferioris. Dorfum nonnihil elevatum.	$\frac{2}{3}''$	$\frac{1}{3}''$	alt 5'''. lat. 18'''
E.	Articulatio epiglottidis parum elevatur et retrorfum trahitur.	Apex ad aciem dentium inferiorum. Dorfum magis elevatum.	$\frac{1}{3}''$	$\frac{1}{6}''$	4''. 18'''.
I.	Eadem mutatio fed major, articulationis major complanatio, epiglottide et limbo glottidis magis elevatis.	Apex ad medium dentium fuperiorum et inferiorum, vel parum int. dent. Dorfum ma ime elevatum. Apex canaliculum format vocem tranfmittens	$\frac{1}{6}''$	$\frac{1}{12}''$	2''. 18'''.
O.	Eadem mutatio fere, quae in A.	Idem fere ltfus, qui in A, ad $\frac{1}{4}''$ magis retractus et elevatus.	$\frac{1}{2}$	$\frac{5}{12}''$	3$\frac{1}{2}$'''. 8'''.
U.	Apertura Epiglottidis et complanatio articulationis paulo minor quam in I. fine notabili elevatione.	Apex paulo magis quam in O. a dentibus inferioribus retractus. Dorfum in parte poftica magis elevatum.	$\frac{1}{3}''$ ad $\frac{5}{12}''$	$\frac{1}{5}''$	2$\frac{1}{3}$'''. 5'''.

Opening measures of the vowel tract [9]

A first reading of the 'Tentamen', especially of §§ 9. and 10. gives the impression that it was written to provide directions for the construction of resonators. In § 9. Kratzenstein assembles a list specifying the aperture dimensions of the vowel tract. These had been determined with the aid of a flat chip of wood which was put upright between the teeth or lips; for particular widths it could then be ascertained which vowels could be correctly articulated by observing the corresponding opening.[10] Kratzenstein points out that these measures are neither exact nor observable simultaneously.[11] In the same context we also find an early mention of the fact that the vowels are involuntarily pronounced with an initial consonant.

The data probably originate from previous works by Kratzenstein, in which he had occupied himself with phonetic as well as anatomical features of the vowels some time before the prize question.[12] In paragraphs 11 to 24 he deals with Denis Dodard and Antoine Ferrein as well as with the advancement of their theses by Albrecht von Haller [13] in order to develop a view of the production of the vowels from a physician's point of view.

Section 2 of the 'Tentamen' is headlined: 'De construendis fistulis, vocales a, e, i, o, u enunciantibus' (about the construction of pipe-whistles which express the vowels a, e, i, o, and u). To begin with, it must be stressed that Kratzenstein did not conceive of the construction of the vowel resonators as an imitation of the vowel tract. Instead, he developed the forms by extensive experiments. It is not surprising that these basically reproduce form principles of the vowel tract, as we will see in the case of 'e'. Their measures, however, are derived acoustically from the sizing of the 'e' to correspond to the tone C' (at that time 247 hertz). For the height of a conus Kratzenstein indicates a length of about 3 'thumbs'[14] (foot/12).

Sections 26 to 30 describe in more or less detail the construction of the resonance bodies prepared for the vowels.[15] This is supplemented by the original drawings displayed in the appendix.

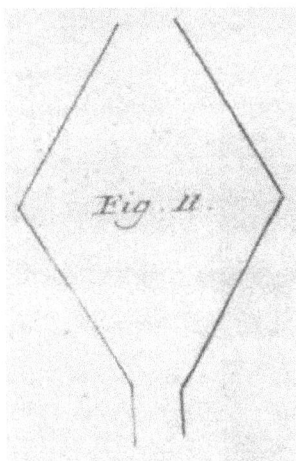

Fig. 11.

Thanks to the alertness of Rüdiger Hoffmann, the reappraisal of Kratzenstein's work including its practical results – the reconstructed instruments – have found a place in Dresden as part of the 'Historische Akustisch-Phonetische Sammlung' (historical acoustic-phonetic collection); this is all the more to be appreciated, since the web page containing my detailed documentation was moved to a somewhat isolated position after my retirement from teaching at the university in 2008.

3. Correcting the drawings

It is advisable not only to view the drawings but to read the descriptions that go with it. The construction of the 'e' is shown in Fig. 11 as well as in the drawing. The explication reads thus: 'On the upper cut-off part sits a tube stub that is wide enough for a little finger to be inserted into it. [16]

As early as 2007 I presumed this to be the correct adjustment as opposed to Fig 11.[17] When the resonator was operated as indicated by the drawing, the audible result was clearly indifferent in comparison to the others. This could be confirmed by measurements: the formants did not resemble those of an 'e' at all. Setting the resonance body against the schematic cuts of the vowel tract indicated the mistake.

If the 'o' is considered as a two tube resonator, the upper one is narrow, while the lower one

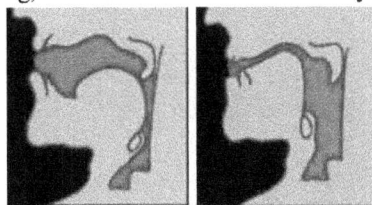

o   e

is clearly broader in comparison. In the case of 'e' the relations are exactly reversed. The mere fact advocates – without reference to the 'Tentamen' – a change in the design of the resonator as shown in the pattern drawing: The acoustic impression as well as the measurement of the formants then produce satisfactory results.

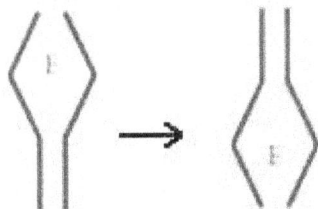

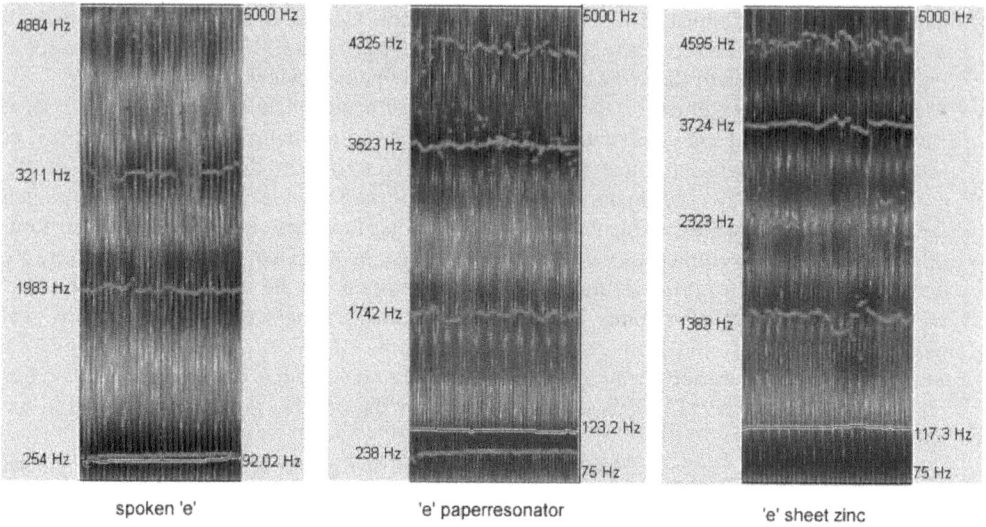

spoken 'e'   'e' paperresonator   'e' sheet zinc

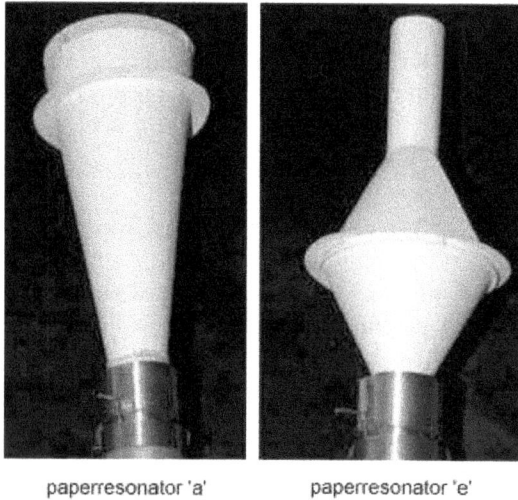

paperresonator 'a'   paperresonator 'e'

4. The i-Problem

In generating the 'i', too, the drawing alone (Fig. 12) is not really helpful. Even though the patterns characterizing the silhouette seem to be largely correct, no vowel resembling an 'i' can be generated with a resonator whose shape follows the pattern. Kratzenstein himself must have been aware of this problem: His article devoted to the 'i' is by far the longest.

As regards the acoustic excitation of the resonators, their description in 'A System of Natural Philosophy', mentioned at the beginning, contains the following explanation: '[Kratzenstein] showed that the sounds of the four vowels, A, E, O, and U, might be obtained by blowing through a reed into several tubes, the forms of which are represented in the annexed figures 1, 2, 3, and 4; and that the sound of I, as pronounced by the French and other continental nations was produced by blowing at a, into the pipe No. 5, without using a reed.' [18] The only

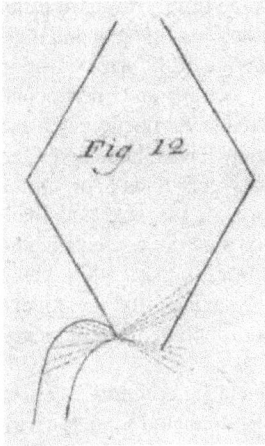
Fig 12.

twentieth century author who appears to have taken notice of it is George A. Miller; he comments it as follows: 'The resonator for i was blown about the opening, the other resonators were activated by introducing a vibrating pipe sheet'. [19]

This explanation, however, is no more than a hint at what the dotted lines to be found in the original drawing (bottom end) can be supposed to signify. Their interpretation is complicated by the incorrect representation of the airstream, for a hollow body blown in this manner – for example a bottle – would be blown above the right edge.

The description in the text explains the underlying principle more clearly: 'Near to this pipe a resonator made out of two truncated cones, as we know them for the vowel 'e', is attached in such a way that the airstream blows at the resonator as is the case with a transverse flute.' [20]

This explanation is a double statement: It points out a method of how to handle this resonator successfully, in contrast to the others. In addition, it intimates that, judged by its construction and functionality from a systematic point of view, the appliance does not, in fact, qualify as a 'resonator'. Kratzenstein is conscious of this and states the reason why his construction lies beyond the scope of the system: 'With regard to this vowel, however, a gap would be left open in our work, and thus I have developed a different kind of flute pipe which generates this vowel with sufficient clarity'[21] While all the other vowel resonators are to be considered bells of reed pipes, Kratzenstein here constructs a labial pipe. Strictly speaking, it cannot, therefore, correctly be called a resonator. Rather, it is a pipe body whose length generates the audible oscillation stimulated by the lip.

All vowel producers were stimulated aerodynamically; this is equally valid for the tongue lamella of a pipe whistle with resonator and the edge of a labial pipe. Until the vibrating air column stands, one hears the stimulating airflow. Hence, Kratzenstein always observes the occurrence of an initial sound 'h' in natural articulation as well as in operating his vowel-tubes. [22]

Aerodynamic processes are left unconsidered in the source-filter model, which is expounded in every phonetics textbook in order to explain the production of vowels. Technically speaking, though, it disregards the same difficulty that Kratzenstein explicated in his work on vowels. In order to close this gap, to my mind a solution should be sought by extending the explication of the vowel production beyond the source-filter theory. Nowhere, however, is such an approach to be found. It would be of interest to work out – at least from a history of science angle – the point in time when investigations in the aerodynamic processes of the vowel production ceased. The electromechanical treatment of the problem, originating in the 19th century, was able to cope without resorting to aerodynamics. I believe, however, this latter field still matters – for the history of science as well as for purposes of vocal education and speech therapy.

5. History of Sciences, practical utility and profit

A reassessment of Kratzenstein's achievements raises a further question, that exceeds the technical aspects of his works. I will outline it briefly. It concerns the applicability of scientific findings and their role in promoting the natural sciences in the eighteenth century.

His invention of a reed pipe with a free reed took up a disproportionate space in subsequent descriptions, and its advantages were praised exceedingly. This was followed by a heated discussion about its authorship in the instrument making craft. For some time, Leonhard Euler had corresponded with Kratzenstein concerning acoustic problems and, as his acquaintance, presumable brought his authority to bear when it came to drawing up the 1780 prize question of the academy. Only two years later, the audience at Kratzenstein's 'lectures on experimental physics' were deeply impressed by his extensive scientific knowledge of phenomena and theories.[23] In 1783, the first manned flight of the 'Montgolfière' was launched, and Kratzenstein quickly responded by publishing his work. 'L'Art de naviguer dans l'Air' in the following year. It contains extensive instructions for the calculation of the size and load-carrying capacity of hot-air balloons of this type as well as deliberations on their scientific and commercial uses. [24] Kratzenstein has been described as a very modest person while being generous towards his employees. This assessment allows us to conclude that he indeed possessed an entrepreneurial mind that was also guided by prospects of income to be achieved by the application of his research.

The origins of modern natural science and the organization of the scientific community are to be sought in the eighteenth century; the impetus of its development is unabated to the present day. It is marked, in the first place, by an increasing systematization of practical knowledge that is further fueled by international exchange. The Petersburg Academy liberally offered free correspondence and publication, which made it extremely attractive for its members. [25] Concurrently, scientific work became increasingly independent of its immediate practical utility, which can be documented by an extensive list of distinct stages characterizing this progress. These practices were taken up by Wilhelm von Humboldt, who incorporated them into a university concept based on his humanistic educational theory.

On closer examination of the historical facts underlying this development, however, the absence of an overall picture becomes obvious. How did the financing and organization of scientific work change up to the status that it reached at the end of the nineteenth century, for example in Germany? When and where and from what motives was this change brought about? It was at this juncture that technical education was integrated into the university system and its working methods. The history of the Dresden University of Technology vividly bears witness to it. It might be rewarding to investigate the development from Kratzenstein's time to the second half of the nineteenth century, in order to critically assess the formation of a corresponding theory.

Its absence is regrettable, the more so as at present we appear to revert to its beginnings – by deeming as progressive the notion that education and research are essentially to be counted as investments.

References

Acknowledgements: Thanks to my colleagues Wilfried Braun and Frederik Heinemann this essay got its English version.

[1]   A System of Natural Philosophy. Philadelphia 1845, p. 265. Thanks to Rüdiger Hoffmann for referring me to this title.

[2]   Korpiun, Christian: 'Kratzenstein wiederbelebt'. 2015. Glottis und Schallquelle in http://www.seam-uni-essen.de/physiologie/akustik/resonatoren/uebersicht_sprechakustik.htm

[3]   Kratzenstein, Christian T: 'Tentamen resolvendi problema ab Academia Scientiarum Imperiali Petropolitana ad annum 1780 publice propositum'. Petersburg 1781. §10, p. 16. http://resolver.sub.uni-goettingen.de/purl?PPN59586435X

[4]   loc. cit.

[5]   Tentamen §§2–8.

[6]   Albrecht von Haller: Anfangsgründe der Phisiologie des menschlichen Körpers. Berlin
        1766. pp. 695 ff.
        http://www.deutschestextarchiv.de/book/view/haller_anfangsgruende02_1762?p=7

[7]   Tentamen §27, p. 39f.

[8]   Korpiun op. cit., 'Kratzenstein wird geadelt'.

[9]   Tentamen §9, p. 15.

[10]   Tentamen, §10, p. 16

[11]   loc. cit.

[12]   Tentamen §§2–8.

[13]   von Haller, Albrecht 1766 pp. 695 ff.

[14]   Tentamen §27, p. 39f.

[15]   Korpiun op. cit., 'Kratzenstein wird geadelt'.

[16]   Tentamen §27, p. 39f.

[17]   Korpiun op. cit. 'Das Problem mit dem 'e''.

[18]   loc. cit.

[19]   Miller, George A.: Wörter. Streifzüge durch die Psycholinguistik Heidelberg, Berlin,
        New York 1992, p. 88

[20]   Tentamen §28, p. 41f.

[21]   Tentamen §28 p. 41.

[22]   Tentamen §8, p. 14.

[23]   Kratzenstein, Christian T: Vorlesungen über die Experimentalphysik. Kopenhagen
        1781. http://reader.digitale-sammlungen.de/resolve/display/bsb10131365.html

[24]   Kratzenstein, Christian T : L'art de naviguer dans l'air. Kopenhagen und Leipzig
        1784. http://reader.digitale-sammlungen.de/resolve/display/bsb10080962.html

[25]   Kopelevič, Judith K.H.: Euler und die Petersburger Akademie der Wissenschaften. pp.
        378f. In: Fellmann, E.A. (Hrsg.): Leonhard Euler 1707–1783. Berlin 2013. pp. 373–
        382. http://link.springer.com/chapter/10.1007%2F978-3-0348-9350-3_19#page-1

Voices for toys – First commercial spin-offs in speech synthesis

R. Hoffmann

Technische Universität Dresden, Institut für Akustik und Sprachkommunikation
ruediger.hoffmann@tu-dresden.de

Abstract: When the collection of phonetic instruments of the Phonetic Institute of the Hamburg University came to the HAPS Dresden in 2005, it included also some small mechanic voices, which can produce single sounds as well as few simple words. These voices are well-known in the phonetic literature as an early attempt to provide hard-hearing people with automatic training tools, following a proposal of the otologist Johannes Kessel in 1899. On the other hand, the phonetic literature never took any notice from the real origin of these interesting pieces. Therefore the author started an investigation some years ago [1], which guided him not only to the interesting field of mechanical voices in the manufacturing of toys and dolls, but moreover back to the roots of mechanical speech synthesis at the end of the 18th century. This paper gives a rough overview about the recent state of this investigation.

1 From Kempelen to Mälzel

Apart from other reasons, the speaking machine of Wolfgang von Kempelen received its fame by a good marketing, which was mainly effected by demonstrations of Kempelen's automata throughout Europe at his journey in the years 1783/84 [2]. The Saxon major-domo Joseph Freiherr zu Racknitz reports about the presentation of the automata in Dresden 1784, that "the speaking machine aroused admiration, while the chess player also produced curiosity" [3]. In 1791, von Kempelen published his summarizing book about the speaking machine [4]. The chess player, however, was stored in the Schönbrunn castle for two decades.

After von Kempelen's death in 1804, the chess player (called "the Turk") came into the ownership of Johann Nepomuk Mälzel (1772–1838). He was a famous German musician, engineer, automata constructor, and entertainer, who is known today mainly as the eponym of the metronome. He restored Kempelen's chess player and demonstrated it together with his own constructions throughout Europe and, from 1825, in America [5].

It is not completely clear whether Mälzel also acquired a copy of the speaking machine from the estate of Kempelen. Anyhow, he came to Vienna already in 1992, where he certainly got in touch with Kempelen and his work. He was educated very well in constructing mechanical musical instruments like the famous "Panharmonicon" (1805) and instrument-playing automata like a spectacular trumpeter (1808). Therefore it appears logically, that he started to equip his automata with voices. As an instance, he presented an automatic tightrope walker, which spoke words like "Oh là là". Best known, the chess player obtained in the winter season 1819/20 the ability to pronounce the word "échec" (check) [6].

Edgar Allan Poe mentions in his famous essay on "Maelzel's Chess-Player" from 1836 [7]: "During the progress of the game, the figure now and then rolls its eyes, as if surveying the board, moves its head, and pronounces the word 'echec' (check) when necessary. [...] The making the Turk pronounce the word 'echec', is an improvement by M. Maelzel. When in possession of Baron Kempelen, the figure indicated a 'check' by rapping on the box with his right hand."

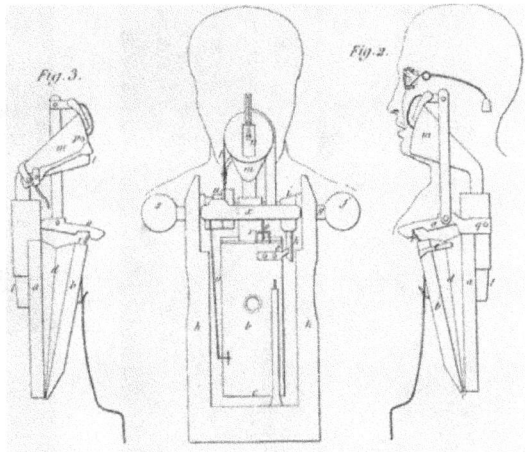

(13)

1600.

31 janvier 1824.

BREVET D'INVENTION DE CINQ ANS,

Pour une mécanique ou automate dite *poupée parlante*, qui prononce, par le jeu de ses bras, les deux mots *papa* , *maman* ,

Au sieur MAELZEL (Jean), mécanicien à Paris.

Figure 1 - Schematic presentation of the Papa/Mama voice from the patent of J. N. Mälzel [8].

2 The invention of the speaking doll in France

Before his relocation into the New World, Mälzel was mainly located in Vienna, where he was appointed to the k. k. Court Chamber Machinist in 1808. But the second center of his life was Paris, where he, for instance, established a factory for the production of metronomes in 1815 [5, p. 140].

At the French Industry Exhibition in Paris 1823, dolls had been shown which pronounced 'Mama', if the right hand was lifted toward the shoulder, and 'Papa', if this was done with the left one [5, p. 184]. The were produced by J. N. Mälzel, who received a patent for this in 1824 (Figure 1) [8].

The patent explains, that the movement of the left arm opens a bellows by means of an eccentric disk. The air from the closing bellows excites a vibrating tongue, the sound of which is modulated by a funnel-like "articulation tract" to produce the vowel *a*. The CVCV structure of the words *mama* and *papa* is simply achieved by closing the funnel twice.

Slightly different to the aforementioned description, the right arm serves mainly for switching between *mama* and *papa*. There is a small hole provided at the funnel. If it is closed, the plosive is produced, otherwise the nasal. We will find the same principle in the construction of Hugo Hölbe below. Some more details have been described and illustrated in [9, p. 325].

We do not know whether real objects from that time have survived. Later, the speaking dolls were equipped with spring drives, which enabled them to kick or to move. Two still existing examples of this type, manufactured by Jules Nicolas Steiner, Paris, in the 1860s, are shown in a catalogue from 1991 [10, no. 167/169].

We want to stress that the Mälzel patent describes the category of "pulling voices", which means that there is a bellows, which must be opened by pulling or a comparable action. The sounds are produced when the bellows is emptied due to the pressure of a spring. There is another, simpler category, which is called "pressure voices", where the user is pressing onto a bellows, which immediately produces a sound. This principle was applied for simple animal voices. The products, which are called "bellows animals" (Balgtiere), are also found from French producers (cf. the example in [11, ch. XIII], Figure 3 a).

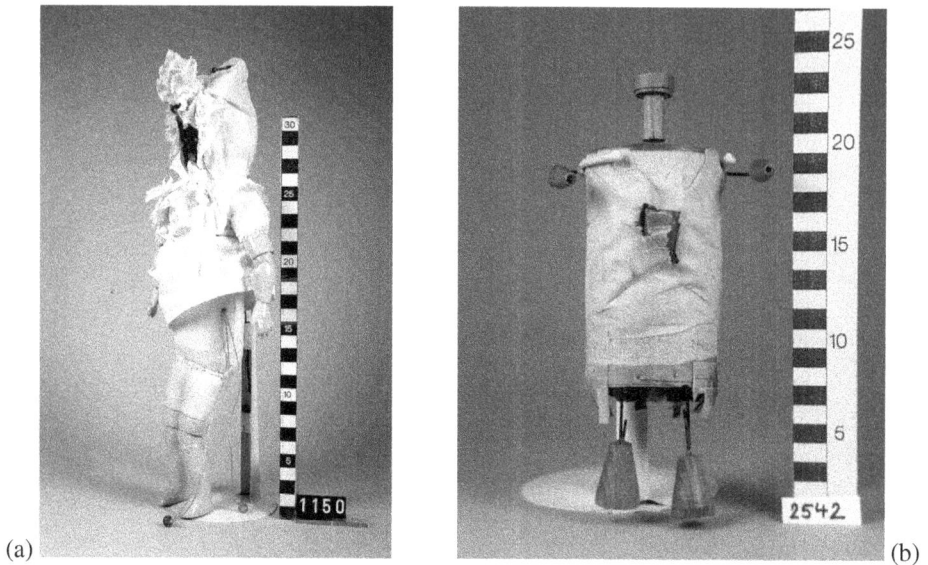

Figure 2 - Exhibits from the Deutsches Spielzeugmuseum Sonneberg: (a) Speaking doll from Carl Bergner; (b) voice of a doll from Hugo Hölbe. Photographs by courtesy of Deutsches Spielzeugmuseum.

3 Voices in the manufacturing of dolls and toys in Germany

3.1 Sonneberg – the world capital of toys

Sonneberg was part of the Thuringian dukedom Sachsen-Meiningen since 1825. The town including the surrounding region formed a traditional location for manufacturing toys, mainly from wood and, following its introduction around 1805, from papier-maché. The production, which was nearly completely performed in outwork and small family-based workshops, developed rapidly over the 19th century, and Sonneberg was called the "world capital of toys" until World War I.

Of course, the invention of the speaking doll influenced the product spectrum [12, p. 232]: "The message of the speaking doll in Paris excited the toy manufacturers in Thuringia. When the first 'täuflinge'[1] were produced in Sonneberg in 1852, the pressed hollow bodies from papier-maché were equipped with the first voices."

Early examples of speaking dolls are generally rare. The Deutsches Spielzeugmuseum Sonneberg preserves an example of the 'täuflinge', shown at [13]. A second speaking doll came to the museum in 1908 (Figure 2 a). Another exhibit (Figure 2 b) demonstrates the speaking mechanism of a doll separately.

Max von Boehn (1860–1932), who was a productive writer of popular works on cultural history, reproduced a photograph of an early "jointed speaking doll" [14, p. 158]. He designates the origin as Bayerisches National-Museum, Munich, but unfortunately this statement proves to be wrong.

---

[1] Täufling (literally: a child to be baptized) is the name of a certain type of dolls.

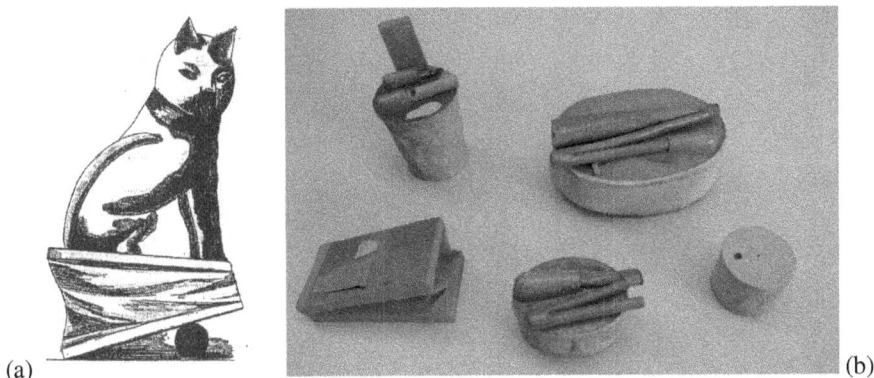

Figure 3 - Examples for pressure voices. (a) Typical construction of a "bellows animal". Copy of an illustration in [11]. (b) Examples of pressure voices. HAPS Dresden, gift of J. and M. Cieslik.

3.2 The voices and their manufacturers

The production of the mechanical voices required special skills. There was a clear differentiation between several jobs in the manufacturing process of dolls and toys [15], and one of the professions was that of the *voice maker*.

Voice maker (Stimmenmacher) was another profession than *bellows maker* (Balgmacher). The latter produced solely pressure voices, which were mainly used in bellows animals, similar to Figure 3 a. A bellows animal was firstly mentioned in a catalogue from 1753/54, and many examples of these nice toys are shown in [16]. Pressure voices were produced in most different versions until the 20th century (Figure 3 b).

In contrast, voice makers produced also the pulling voices for dolls and animals, which were more complicated and showed high variability. Among them were also monstrous constructions for big animals in store windows and similar displays (Figure 4 a). The life of the voice makers was described in the ethnographic literature at different places, for instance [17].

It is reported in [12, p. 232], that the mechanic Hensold in Neustadt was the first manufacturer of voices for dolls in that area. However, the first voices were not articulated, and it lasted until 1857/58, when the first voices in Sonneberg were able to pronounce *Mama* and *Papa*. The modeler Christoph Motschmann from Sonneberg received a patent for a *Mama / Papa* voice from the Ministry in Meiningen on April 30, 1857, which was published in the relevant law gazette at May 6 (cited in [12, p. 201]).

The workshops of the voice makers were distributed among different places in the Sonneberg area. The most detailed examination of the structure of the working world was done for the village Judenbach [16]. Regarding the town of Sonneberg, we know eight voice makers according to the address book from 1911. The most important among them was Hugo Hölbe.

3.3 The voice maker Hugo Hölbe

Hugo Hölbe (March 25, 1844–May 24, 1931) is mentioned as a voice maker in the address books of Sonneberg from 1887 to 1911 [18]. His work forms the most extensive stock of historic voices in the Deutsches Spielzeugmuseum Sonneberg. He was since 1897 a member of the Board and the Committee of the Industrieschule, the collection of which formed the basis of the recent museum. The main part of the voices gave Hölbe to the museum as a gift [19]. Interestingly, he also donated a representative selection of five pulling and three pressure voices

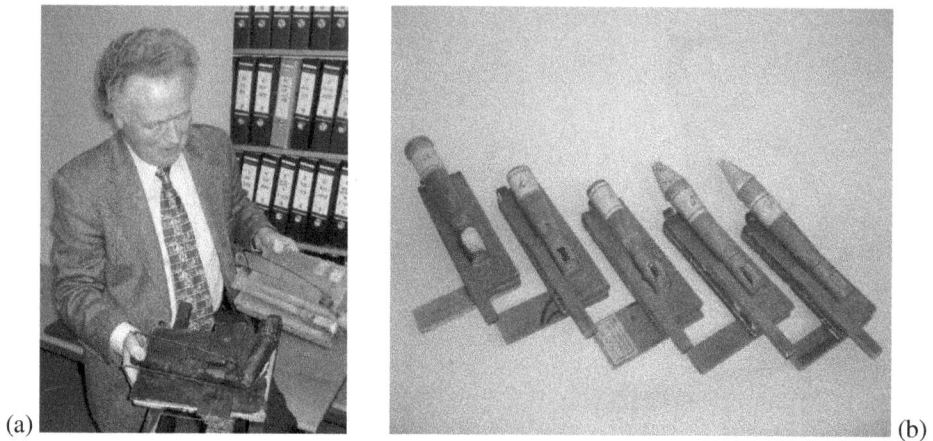

(a)   (b)

Figure 4 - Pulling voices from the estate of Hugo Hölbe in the collections of the Deutsches Spielzeug-museum Sonneberg. (a) Dieter Mehnert demonstrates two of the monstrous animal voices. (b) Voices for the vowels *a, e, i, o, u*. Photographs from a visit in February 2007.

to the Deutsches Museum Munich in 1909 [20].

Basing on the estate in the Sonneberg museum, we have a good overview on the spectrum of Hölbe's products. The predominant part of about 33 objects includes animal voices from the frog to the elephant and, of course, nearly all domestic animals. More interesting in the focus of this paper (which is still synthetic speech) is a set of voices, which can produce the vowels *a, e, i, o*, and *u* (Figure 4 b), the diphthong *au*, and the consonant *r*. According to the files of the museum, the pieces are manufactured around the year 1870.

Moreover, the collection includes also voices, which are able to produce small words. We know the principle from Mältzel's voice for *Papa / Mama*: When the bellows is closing, its movement does not only press the air in the "articulation tract", but controls also some mechanic parts, which, for instance, close the "mouth" of the device temporarily. From the mechanical point of view, a kind of link motion control (known in German as Kulissensteuerung) is implemented. Apart from the classics *Papa / Mama*, the Sonneberg collection includes voices for *Maria, Emma, Mimi, Anni*, and finally, as Hölbe's masterpiece, for *Hurra* (or *Hurrah*, in older German writing).

The *Hurra* voice (Figure 5 b) forms the attraction of any collection of pulling voices. The production of a clearly recognizable *r*-sound was the most essential prerequisite. This problem was solved by mounting a small flap over the end of a simple voice (Figure 5 a), which really produces a sound like the purring of a cat. Hölbe was obviously proud of his invention and registered it as protected design in 1880.

Hugo Hölbe was also holding a patent from 1883 (DRP 26 082) for a singing mechanism for dolls, where a bellows drives a kind of musical roller. In 1923, he registered a "voice for dolls and the like" as protected design (DRGM 847 182) [12, p. 121].

It is worth to mention that there was another voice maker Hölbe with the Christian name Richard. He worked in Oberlind, which is today a district of Sonneberg. He was holding several patents [12]. One of them (DRP 62 868 from 1890) comes back to the switching act between *Papa* and *Mama* in Mältzel's invention, which was performed be closing and opening a hole, respectively. Hölbe invented a self-acting mechanism, which alternately closed and opened the hole, causing the doll to switch between *Papa* and *Mama* automatically.

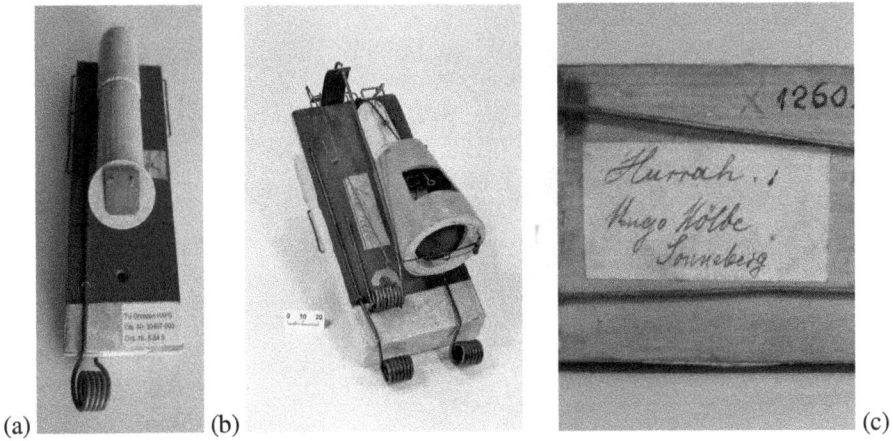

(a) (b) (c)

Figure 5 - The *Hurra* voice from Hugo Hölbe. (a) Voice for producing the consonant *r* separately. HAPS Dresden. (b) Voice for producing the word *Hurra*. HAPS Dresden, photograph by Rolf Dietzel. (c) Signature of Hölbe at the underside of the *Hurra* voice in the collection of the Deutsches Spielzeugmuseum Sonneberg. Photograph from a visit in February 2007.

4 The proposal by Johannes Kessel

The voices from Hölbe found an exceptionally application in the field of pronunciation training. Johannes Kessel (1839–1907) was an extraordinary professor of otology at the Jena university since 1886[2]. It was an element of his tasks since 1888, to supervise the pupils of the institute for blind and deaf-and-dumb children in the neighbored residence town Weimar. Therefore, Kessel was engaged in the treatment of deaf-and-dumb people in growing extent.

In that time, deaf-and-dumb people were frequently considered as mentally deficient. Methods for improving their education were developed step by step. A big common meeting of the otologists and the deaf teachers, which took place in Munich in September 1899, can be considered as a milestone. The treatment of probands who have a certain residual of hearing capacity formed a controversial point at the meeting. It sounds very modern when Kessel pointed out in his contribution [21], that these patients need very much time for pronunciation training, which are not enough teachers are available (and payable) for. Therefore he proposed, that the patients should be equipped with automatic training tools.

Kessel demonstrated in Munich a set of mechanical voices, which included at least the vowels and the words *Mama*, *Papa*, and *Marie*. He mentioned in his contribution explicitly, that they were manufactured by Hugo Hölbe in Sonneberg. Unfortunately, we know nothing about the way, in which Kessel came in touch with Hölbe. Maybe, he bought a copy of the "speaking picture books", which we will describe in Section 6, for one of his four children and learned about the mechanic voices in this way.

There seemed to be some understandable criticism at the conference regarding the poor quality of the synthesized speech [22]. Kessel was even ridiculed in a journal for his "shrieking dolls" [23]. More than one century later, we know that the development of synthetic speech required a long way to achieve a more or less satisfactory quality level. Nowadays, the computer-aided pronunciation training (CAPT) is one of the most emerging application fields in speech technol-

---

[2]Cf. the remark on Kessel and his detailed biography in Section 4.2 of the contribution of Hoffmann & Mehnert in this volume.

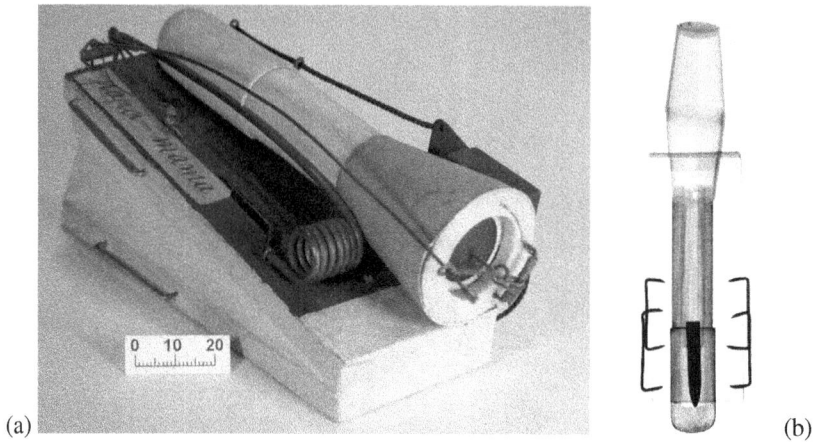

(a)　　　　　　　　　　　　　　　　　　　　　　　　　　(b)

Figure 6 - Examples for the voices in the HAPS collection. (a) The voice for *Mama/Papa* with opened bellows. Photograph by Rolf Dietzel. (b) X-ray photograph (top view) of the voice for the vowel *o*.

ogy not only for second language learners, but also in rehabilitation engineering. We appreciate Kessel as a pioneer in this field, who recognized the potential of synthetic speech[3].

Hermann Gutzmann sen. (1865–1922) was probably the first who recognized the importance of Kessel's proposal. His father Albert Gutzmann (1837–1910) had joined the Munich meeting as Director of the Berlin institute for the deaf-and-dumb. Hermann Gutzmann, who was the founder of the Phonetic Laboratory of the Berlin University, inserted a reference to the voices from Kessel in the second edition of his *Speech therapy* [24].

5　The mechanical voices in the HAPS Dresden

Giulio Panconcelli-Calzia (1878–1966), the Director of the Phonetic Laboratory (later Institute) of the Hamburg University, was not only an important representative of experimental phonetics, but also author of numerous historic publications. Basing on the remark of Gutzmann, he mentioned Kessel's voices in his historic work and expressed [25] [26, p. 48]: "The original equipment is situated in the Phonetic Laboratory in Hamburg." When the purchase book of the laboratory was laid out in 1913, this set of "artificial voices" was already among the inventory, labeled with a value of totally 66 Reichsmark.

The hint from Gutzmann and Panconcelli-Calzia to Kessel's voices was cited in the phonetic literature from time to time, but neither Panconcelli-Calzia nor other authors discussed the real origin of the mechanic synthesizers.

The voices came with all other exhibits from Hamburg to the historic acoustic-phonetic collection (HAPS) of the TU Dresden in 2005. The set includes ten pieces: 5 vowels, *au, r, Papa/Mama, Emma, Hurra*. Examples are shown in Figures 5 a, 5 b, 6 a and in the catalogue of the HAPS [27, p. 186]. Due to kind support of the laboratories of the department, X-ray photographs of the objects were produced (Figure 6 b).

If these voices are compared to other work from Hölbe (like that in Figure 4), there is some doubt whether they are really the originals from Kessel and Hölbe. Later on, the precision engineer Julius Ganske from Zehlendorf (today part of Berlin) offered replicas of the voices,

---

[3]Kempelen also explained in his book about the speaking machine, that the main usage might be in teaching deaf people and healing patients with wrong pronunciation [4, p. XI].

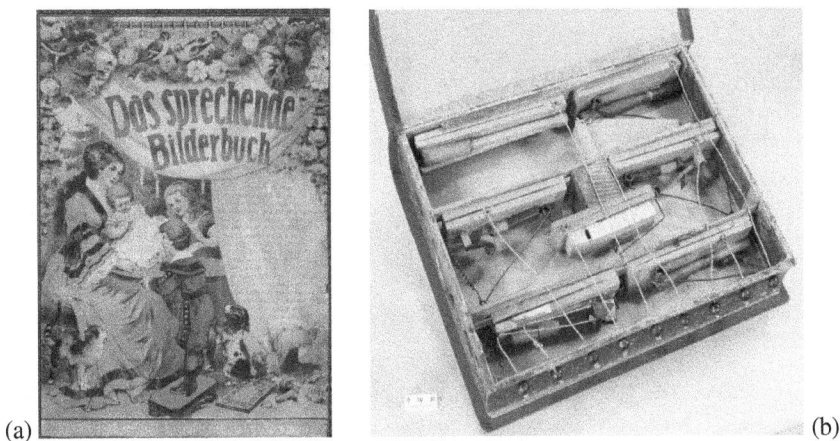

(a)   (b)

Figure 7 - The speaking picture book. (a) Cover of the German edition. (b) View of the (opened) hollow
space, which contains the voices. Photographs of the copy from the 16th edition in the HAPS, Dresden.

the appearance of which is very close to that of the voices in the HAPS [28]. We simply suffer
from a lack of information, whether Panconcelli-Calzia got the collection from Kessel or from
Hölbe or from Ganske.

Ganske included the voices in his catalogue of devices for the experimental phonetics as "artificial vowels and words for the explanation of the effect of the vowel tract". The set corresponds
to the aforementioned, except from the *au*.

The former Hamburg, now HAPS, collection includes another set of 12 voices in a display box,
which undoubtedly goes back to Hölbe [27, p. 185]. Dated on August 22, 1917, the purchase
book of the Phonetic Laboratory Hamburg registers the purchase of 12 "voice mechanisms"
(5 vowels, *au, r, Hurra, Papa, Marie, Emma*, cat voice) for 23 Reichsmark from Hugo Hölbe,
Sonneberg. The inventory number is still readable and allows a unique identification.

6   The speaking picture book

The mechanical voices were applied in the first book with multimodal properties. The book-
seller and publisher Theodor Brand (* 1847) from Sonneberg developed the idea for a "speaking
picture book" in 1874 and received a patent for this in 1878 [29, 30].

At a first glance, the speaking book looks like a regular book in quarto size (Figure 7 a). The
book block, however, turns out to be a box containing a set of pulling voices (Figure 7 b). Some
book pages are mounted between the book cover and the box, showing different animals or
people along with correspondig texts or poems. A knob at the long side of the box is assigned
to each picture by a printed arrow. Pulling the knob activates the voice of the animal. The effect
is really amusing, and the book is now a desired object for collectors.

Theodor Brand published the book in different versions over a long time. It achieved worldwide
distribution, because the text pages were printed in several languages (German, English, French,
Spanish). Therefore it was a big commercial success, especially at the US market [31].

The success of Brand's book inspired more manufacturers at the begin of 20th century, therefore other designs than that from Figure 7 are also found. However, the production of speaking
picture books with the complicated pulling voices remained restricted to Sonneberg and Judenbach. At the latter location, Albin Matthäi (1892–1974) should be mentioned, because he

Figure 8 - Examples for the later, simplified "speaking picture books" with pressure voices. (a) Book with seven double-sided cardboards from the manufacturer AMA, Judenbach, probably 1930/40s. (b) Book with German and French text, unspecified, probably around 1960. Private collection.

introduced books with pressure voices in his company "AMA" around the year 1930. These cheaper books (Figure 8 a) were still produced after World War II in the GDR. More details can be found in [16, pp. 251–254].

Speaking picture books with pressure voices consist of several cardboards with pictures of animals. Flat special versions of pressure voices are inserted into the cardboards. The sound is produced, when the user is pressing a picture. Although the term "speaking book" was furthermore used for the product, the simple sounds have nothing to do with speech; actually it is a "screeching book". Interestingly, the book type survived somewhere, as we learn from antiquarian offers like that in Figure 8 b. Unfortunately, these books have no imprints, but the design points to the 1960s.

7 Conclusion

We followed the development of the mechanic voices from the late Baroque until the 1960s, spanning nearly two centuries. It is interesting that the success story lasted as long, despite of the invention of the phonograph by Edison in 1877/78. Even Panconcelli-Calzia wondered about the fact, that Kessel made his proposal using the mechanic voices in 1899, when the phonograph was more than ten years old [25]! It should be also remembered, that the invention of Edison enabled the production of a new generation of speaking dolls in the US (Figure 9), which surprisingly did not win the competition to the dolls with mechanic voices. "Phonograph dolls" have also been produced in Germany [12, p. 277] and France [10, p. 161]. Later on, the application of gramophone discs for speaking dolls was more successful. They were also produced in Sonneberg [16, p. 152]. But all these inventions disappeared in the history after the introduction of electronic systems.

Acknowledgement

The author thanks Dipl.-Hist. Sonja Gürtler from the Deutsches Spielzeugmuseum Sonneberg for giving access to the exhibits of the museum and for useful hints. Thanks also to Dr. Ernst Hofmann, the former Director of the museum, who did a lot of research on the history of the speaking picture books. He kindly submitted an unpublished text [29] with valuable information.

Figure 9 - Speaking doll (right) and small phonograph (left) from Edison. The central picture shows how the recording on the wax cylinders was performed. Details of a contemporary newspaper illustration.

References

[1] Hoffmann, R.; Mehnert, D.: Die Kesselschen Stimm-Mechaniken in der historischen akustisch-phonetischen Sammlung der TU Dresden. In: DAGA 2007, Stuttgart, 19.–22. 3. 2007, Tagungsband "Fortschritte der Akustik", 401–402.

[2] Reininger, A.: Wolfgang von Kempelen – eine Biografie. Wien: Praesens Verlag 2007.

[3] Racknitz, J. F.: Über den Schachspieler des Herrn von Kempelen und dessen Nachbildung. Leipzig und Dresden: Breitkopf 1789.

[4] Kempelen, W. v.: Mechanismus der menschlichen Sprache nebst der Beschreibung seiner sprechenden Maschine. Wien: J. B. Degen 1791.

[5] Leonhardt, H.: Der Taktmesser. Johann Nepomuk Mälzel – Ein lückenhafter Lebenslauf. Hamburg: Kellner Verlag 1990.

[6] Standage, T.: The mechanical Turk – the true story of the chess-playing machine that fooled the world. London: Allen Lane The Penguin Press 2002.

[7] Poe, E. A.: Maelzel's Chess-Player. In: The Southern Litarary Messenger, April 1836. Online version (12. 7. 2015): http://xroads.virginia.edu/~hyper/POE/maelzel.html

[8] Maelzel, J.: Mecanique ou automate dite poupée parlante [...]. French Patent No. 1600, 31. 1. 1824.

[9] Chapuis, A.; Droz, E.: Automata – a historical and technological study. English translation. Neuchatel: Éditions du Griffon 1958.

[10] Krafft, B.; Bordeau, A. (Eds.): Traumwelt der Puppen. Munich: Hirmer 1991.

[11] d'Allemagne, H. R.: Histoire des Jouets. Paris: Librairie Hachette & Cie. 1903.

[12] Cieslik, J. and M.: Cieslik's Lexikon der deutschen Puppenindustrie. Jülich: M. Cieslik Verlag 1984; 2nd, revised ed. 1989.

[13] Deutsches Spielzeugmuseum Sonneberg, museum-digital (17. 7. 2015): `http://www.museum-digital.de/thue/index.php?t=objekt&suinin=56&suinsa=215&oges=2682`

[14] Boehn, M. v.: Dolls and puppets. Translated by J. Nicoll. Boston: Branford, revised edition 1956. [German original published in 2 volumes, Munich 1929.]

[15] Meyfarth, B.: Die Sonneberger Spielzeugmacher. Zu den Arbeits- und Lebensbedingungen der Sonneberger Spielzeugmacher Ende des 19. und Anfang des 20. Jahrhunderts. Schriftenreihe des Spielzeugmuseums Sonneberg, 1981.

[16] Hahn, R. and O.: Sonneberger Spielzeug – Made in Judenbach. 300 Jahre Spielzeugherstellung an der alten Handelsstraße. Münster etc.: Waxmann 2010.

[17] Zien, G. (Ed.): Von Puppen, Griffeln und Kuckuckspfeifen. Aus der Arbeitswelt unserer Großeltern. Sonneberg: Landratsamt 1995.

[18] Message from the Stadtarchiv Sonneberg, November 27, 2006.

[19] Message from the Deutsches Spielzeugmuseum Sonneberg, December 6, 2006.

[20] Message from the Deutsches Museum München, 2013.

[21] Kessel, J.: Demonstration von Apparaten zur Erzeugung künstlicher Laute. In: Bezold; Passow (Ed.): Verhandlungen der Versammlung Deutscher Ohrenärzte und Taubstummenlehrer in München am 16. September 1899. Berlin: E. Staude 1900, 28–29.

[22] Denker, A.: Bericht über die Versammlung deutscher Ohrenärzte und Taubstummenlehrer zu München. Archiv für Ohrenheilkunde 47 (1899) 3, 189–208.

[23] Footnote in: Blätter für Taubstummenbildung 14 (1901) 10, 156–157.

[24] Gutzmann, H.: Sprachheilkunde. Vorlesungen über die Störungen der Sprache mit besonderer Berücksichtigung der Therapie. 2nd edition, Berlin 1912.

[25] Panconcelli-Calzia, G.: Die Sprechmaschinen unserer Ahnen. Wissen und Fortschritt 4 (1930) 5, 132–136.

[26] Panconcelli-Calzia, G.: Quellenatlas zur Geschichte der Phonetik. Hamburg: Hansischer Gildenverlag 1940.

[27] Mehnert, D.: Historische phonetische Geräte. Katalog der historischen akustisch-phonetischen Sammlung (HAPS) der Technischen Universität Dresden, erster Teil. Dresden: TUDpress 2012.

[28] Ganske, J.: Illustriertes Verzeichnis von Apparaten für die experimentelle Phonetik. Berlin: Verlagsdruckerei Zehlendorf, s. a. [between 1914 and 1921]

[29] Hofmann, E.: Das "Sprechende Bilderbuch" – Ein Welterfolg aus Sonneberg. Manuscript of a lecture, unpublished, submitted 26. 5. 2015.

[30] Cieslik, J. and M.: Das sprechende Bilderbuch – Eine Sonneberger Erfindung. Cieslik's Puppenmagazin (1991) 1, 46–48.

[31] Speaking Picture Book. In: Scientific American. A weekly journal of practical information, art, science, mechanics, chemistry and manufacturers. New Series, vol. XLII, No. 12, New York, March 20, 1880, p. 179.

The contribution of the kymograph to the description of African languages

Didier Demolin
Laboratoire de Phonétique et Phonologie, CNRS/Sorbonne Paris Cité
ddemolin@univ-paris3.fr

The kymograph, one of the main devices used in early experimental phonetics, was quickly exploited to describe the sounds of languages in sub-Saharan Africa. By 1926, Doke used this device in its phonetics Zulu, to objectify the description of the main sounds of this language: vowels with epiglottal friction; the difference between aspirated and non-aspirated explosive consonants and consonants he describes as ejectives explosives; the three types alveolar laterals (voiced, voiceless and voiceless lateral); the duration of vowels and the different levels of tones. Doke shows that these are the plots of kymograph which to observe a similarity between the consonant bilabial implosive [ɓ] and clicks. The kymograph plots, presented in Figure 1 show indeed clearly a negative airflow at the release of the implosive and of the dental click contrasting with the positive flow observed after a bilabial explosive consonant (Doke 1926: 60).

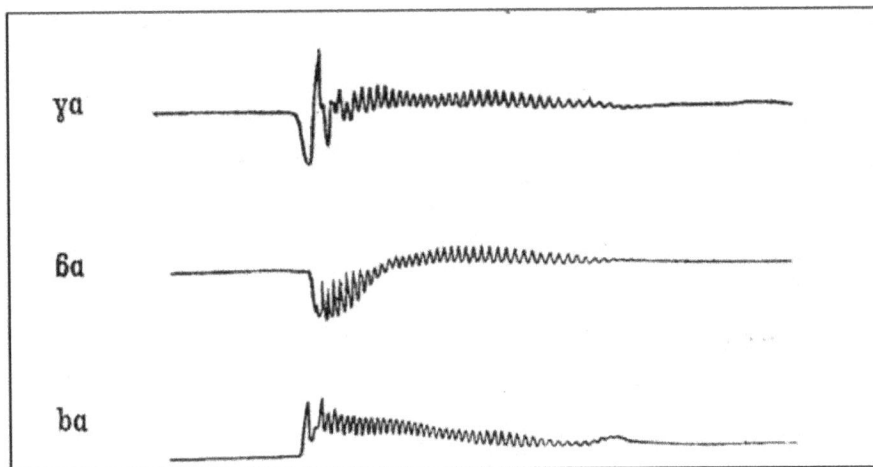

Figure 1. Kymograph plots of Zulu showing the airflow at the release of a click [ɣa] described as dental voiced by Doke (1926), a bilabial implosive [ɓa] and explosive [ba]. (The symbol of the click is now [ǀ] in the transcription of the International Phonetic Association), Doke (1926).

The kymograph was also employed by Westerman and Ward (1933) to show the difference between explosives and implosives in Igbo and Efik. Besides the difference between the two types of bilabials consonants, already highlighted by Doke, they provide examples of the labio-velar consonants [kp] for which there can be a negative flow upon release. These data show a difference between the release of the consonant [kp] in Efik, which is followed by a positive airflow, and Igbo that does not show the positive flow and is described as implosive.

Doke (1931) in his remarkable study on the comparative phonetics of Shona still makes a systematic use of the kymograph to describe the consonants of this language. The combination of nasal and buccals plots or of those with and laryngeal vibration plots allows it to establish the presence or absence of voicing and the boundary between the nasal and oral parts in complex consonants.

A document, unpublished and very interesting (provided by the late Anthony Traill), where Pienaar uses the kymograph, discusses the types of release of the Zulu clicks (Figure 2). Pienaar was a professor of speech therapy at the University of Witwatersrand in the years 1930-1940. He interacted and worked with Doke and created a phonetics laboratory in which students received training techniques, which would be called instrumental phonetics today. Doke and Pienaar would have used the material to make its measures on the airflow of clicks when working with Zulu and Bushman. In his notes, Pienaar describes three phases in the production of clicks: the production, the acoustic result of the anterior release and what he called other members of the click's composition. For Pienaar, clicks are actually compounds phonemes consisting in one or more acoustically different speech sounds to be evaluated 'monophonically'. The first part of the phoneme is the actual suction, while the second may have different acoustic qualities other than the sounds of relaxation and friction. The sound of the suction release is generally described as the click itself. It is usually identified from where the anterior release was made. Pienaar describes the following types of clicks: (1) bilabial or (2) a labio-dental variant; (3) inter-dental, a variant of (4) dental; (5) alveolar; (6) prepalatal; (7); alveolar or prepalatal as (5 and 6), but with a lateral release and (8) retroflex. Each of these types of clicks can be followed in various ways to release the velar closure. These are illustrated in Figure 2 which shows, combining oral flow plots (M) and the vibration of the larynx (L): a voiceless velar stop (kymograph plot –from now on Tk-I); an aspirated voiceless velar stop (Tk II); a voiceless velar affricate (Tk III); a voiced velar stop semi-plosive (Tk IV and V); a voiceless velar stop semi-plosive (Tk VI); a velar ejective (Tk IX); a velar affricate ejective (Tk X). The plots (Tk VII) (Tk XI) show clicks accompanied by a glottal fricative and a glottal stop. Table 1 summarizes the data.

Table 1. Words analyzed by Pienaar in his study of the posterior releases of Zulu clicks. The numbers correspond to those in Figure 2.

(25)	i|ki	(23)	i|$^{?}$i
(26)	i|k^{h}i	(30)	i|k$^{?}$i
(27)	i|kxi	(31)	i|kx^{h}i
	i|gi	(29)	i|gi
(28)	i|gi	(24)	i|^{h}i

Piennar documents are among the first to demonstrate the presence of the two parts of clicks on an experimental basis. His study of the types of posterior releases shows a remarkable understanding of the phenomenon and prefigures the concerns of contemporary researchers.

Beach (1936) in its phonetics of the Hottentot (Nama) language emphasizes the usefulness of the kymograph to study the problems of nasalization, duration and intonation. Some selected plots clearly highlight the clicks of the language and the issues discussed in his treatise.

The identification of tones is fundamental to understand the phonetics of Sub-Saharan languages. The description of this phenomenon took some time before being controlled and clearly illustrates the difficulties researchers had to describe the phenomenon experimentally, in a suitable framework. The famous Laman study (1922) on the musical accent and intonation in Kongo is one of the first systematic work on tone. The satisfactory instrumental treatment of tones, easy as it may seem, has been realized in the second half of the twentieth century. The main problem was to clearly identify the categories of tones -the tonemes- and then process their achievements and phonetic variations. It seems that the first to use the term toneme treatment Beach (1924), as emphasized by Doke (1926: 198).

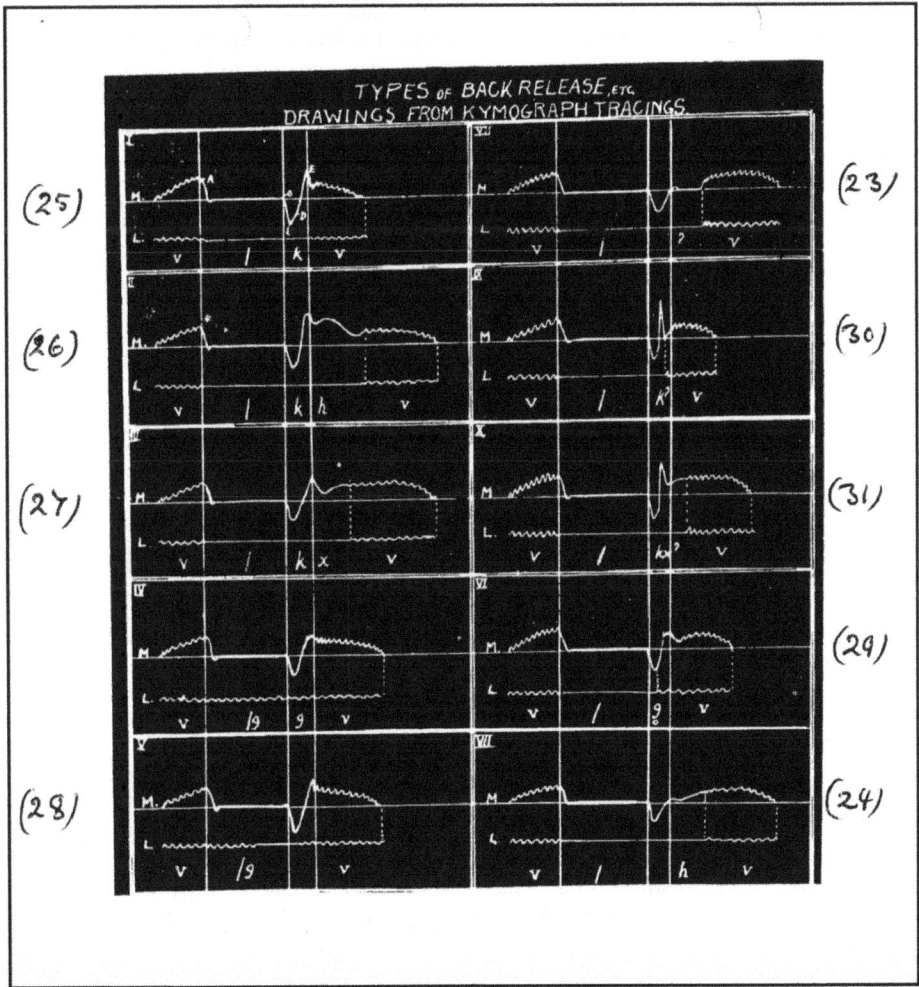

Figure 2. Unpublished kymograph plots by Pienaar describing the types of posterior release of click for a dental click in Zulu. Unpublished document provided by Anthony Traill.

The problem may seem trivial today, but it illustrates the relationship between the abstract aspects (phonological) and concrete (phonetic) in the study of speech. Even if the processing of tones by Doke (1926) and Beach (1936) with the kymograph has only historical interest, it is worth to describe it. This to understand the conceptual difficulties that the researchers had in the early twentieth century, when working on tone languages.

Doke distinguished what he called the characteristic and significant tones. He viewed the former as a method to group the succession of musical heights characterizing a particular language or group of languages. The latter held an active role in grammatical meaning of language and were a way to distinguish different meanings of the words that were phonetically similar. Doke identified 9 tones in Zulu. We should not to believe that his understanding of tone was naive, as shown in the following statement: "*Only a comprehensive study to discover the true phonemes could reduce my system to less than 9 tones for recording*" (Doke 1926: 199). Doke adds that it is not the absolute height that counts, but the relative height, the extent of intervals. To measure 9 tones he identified in Zulu, he used the following method: a kymograph recorded the trace of an electric tuning fork that produced 100 vibrations per second. A native speaker, in a mask connected to a kymograph, then produced the 9 tones. The plots of these recordings were then placed below those of the tuning fork. The number of vibrations of each of these 'tonal curves' corresponding to the length of 100 vibration of the tuning fork was then counted and the results compared to the table of notes of the chromatic scale of Rayleigh (1877). Figure 3 illustrates the result of this process. To transcribe the tones in relation to a text, he then used a number system, which ranged from 1 (highest tone) to 9 (lowest tone). The modulated tones were transcribed with a line between the two levels.

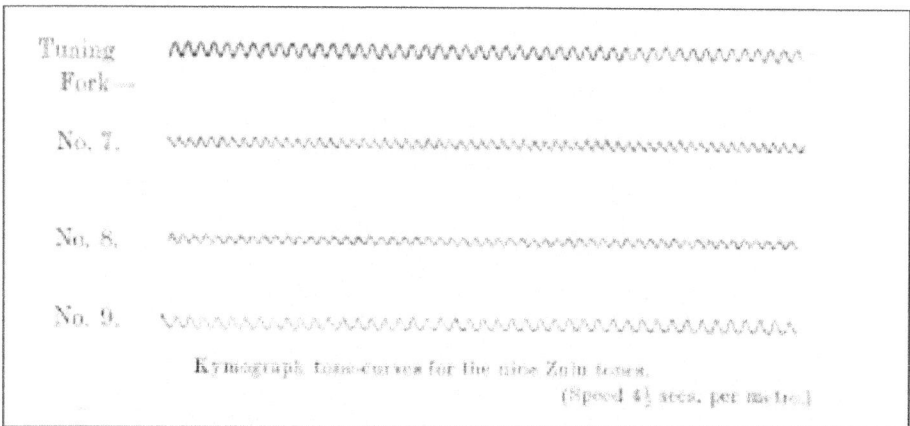

Figure 3. Kymograph plots of tone curves of 3 of the 9 tones identified by Doke in Zulu. The tones are measured in vibrations per second; tone 7 = 116.25, tone 8 = 107, tone 9 = 89. The three tones correspond to the rating A#, Ab and F+, on the Rayleigh (1877) scale, Doke (1926)

Beach (1936) also used the kymograph as a support to the study of tones. This topic is also a highly developed phonetic point of the Hottentot. However, no plot is given in his description, he mentions that it is because of cost of publication and space prevented him from publishing these kymograph plots. Anyway, Beach methodology is interesting to describe, for it differs from that of Doke and because it shows the concerns of these researchers to put the experimental data in an appropriate conceptual framework. Data were recorded on an Edison phonograph. Recordings were standardized to the phonetic analysis using a tuning fork that Beach called again philharmonic 3. (Here Beach probably refers to the 439 Hz as it was defined by the Royal Philharmonic Society ' in 1896, since it was not until 1939 that the International Association stared the measure, known as the concert A, at 440 Hz). At the beginning of each record, before the speaker told a Hottentot story, Beach made the tuning fork vibrate in the phonograph and also sang an A in the recorder. This A was the reference height. Subsequently, when records were used on different phonographs, for tonetic analyzes, the reference height reproduced at the beginning of the recordings was tested against a tuning fork giving the Philharmonic A. If the heights of the recording and the pitch were different from each other, the speed of the phonograph was then adjusted until the reference height matched the tuning fork. When this was done, the tone of the record was identical to that of the speaker when the recording was made. Beach then mentions the existence of devices for the automatic transfer of the gramophone curves to smoked paper, in order to realize the analysis of tone curves with mathematical accuracy. Beach, for reasons of time, apparently did not made much use of the kymograph plots for his tones transcriptions. He used what he calls 'the method of the ear' for this, that is to say, an auditive transcription. Beach transcriptions were then set on a musical score making reference to the 7 types of tones he identified for Hottentot. This is illustrated in Figure 4, which shows the transcription of a piece of text, and the reference heights identified by Beach, for tones in Hottentot.

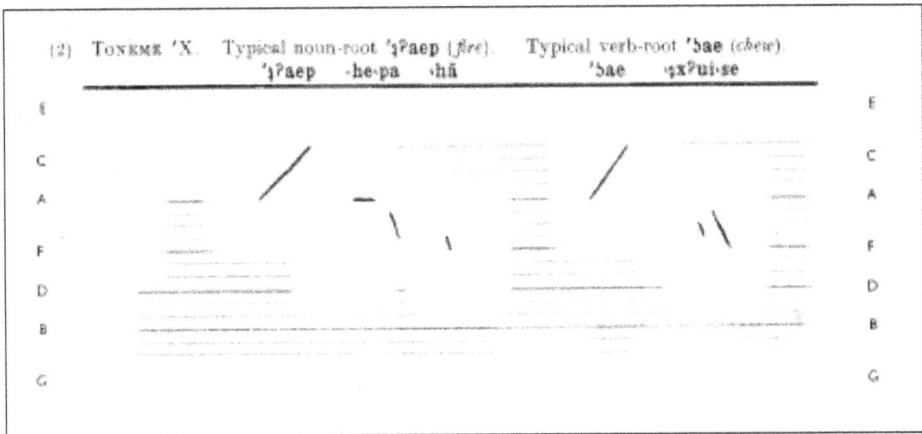

Figure 4. Transcription of the toneme X (rising high) identified in the Korana dialect of Nama by Beach. The Reference lines of the musical score correspond to the notes (G), (B), (D), (F), (A), C (C) and (E). The transcription shows a nominal root and a verbal root starting by the toneme 'X, Beach (1936)

The way Beach transcribed tones is also interesting because it allows to introduce the discussion on the issue of identification of what would now be called phonological tones. This is illustrated in the criticism Beach (1936) addresses to Doke (1926) and Tucker (1929) who are criticized for their misuse of the term toneme. For Beach (1936: 127) "A tone marks system must have as a basis the same principle as the phone system marks (phonetic symbols). In the same way that the phonetic transcription depends on the phonemic principle the tonetic transcription should depend on the tonemic principle ".

Doke (1926: 198), meanwhile, distinguished between: phonetics that deals with phones, groups of phones and phonemes and tonetic that handles tones, tonal groups and tonemes. It is not difficult to realize that the opposition between Doke and Beach come from the fact that the second identifies a phonological level, where the former does not. Doke, and to some extent Tucker faced representational level problems (phonetics, phonology, morphophonological) that were not properly identified in the Bantu languages and still pose problems for bantu specialists. Again, it will take decades to realize that a number of pitch variations are not significant but are linguistically due to mechanical linkages that occur in the process of speech production (Ladefoged 1968).

References

Beach, D.M., 1924, «The science of tonetics and its application to Bantu languages», *Bantu studies*, vol ii, n° 2, p. 75-106.

Beach, D.M., 1938, «*Phonetics of the Hottentot Language*», Cambridge, Heffers.

Doke, C.M., 1926, «The Phonetics of the Zulu language», *Bantu studies,* vol II, Special Number.

Doke, C.M, 1931, «*A Comparative Study in Shona Phonetics*», Johannesburg, University of Witwatersrand Press.

Ladefoged, P., 1968, «*A Phonetic Study of West African Languages*», Cambridge, Cambridge University Press.

Laman, K. E., 1922, «*The musical accent and intonation in the Kongo Language*», Stockholm.

Rayleigh, J. W. S., 1877, «*The Theory of Sound*», vol. 1, MacMillan, London.

Tucker, A., 1929, «The Comparative Phonetics of the Suto-Chuana Group of Bantu Languages», London, Longmans, Green and Co.

Westerman, D. et I. Ward, 1933, «*Practical phonetics for students of African Languages*», Oxford, Oxford University Press.

A 75-year-old Hungarian spontaneous speech database

M. Gósy

Research Institute for Linguistics, Hungarian Academy of Sciences
gosy.maria@nytud.mta.hu

Abstract: The first attempt to develop a large collection of recorded speech material in Hungarian was made by the phonetician Lajos Hegedűs in the 1940s. He wanted to preserve the sounding of the various Hungarian dialects in the country and even outside Hungary with the purpose of analyzing, among other things, the intonation, pauses and rhythm of speech in those dialects. This paper introduces the properties of the Hegedűs Archives on the one hand, and discusses the results of investigations that aimed to compare the speaking strategies of people that could be the great-grandparents of the speakers living now, on the other hand.

1. Introduction

The proverb *What is spoken flies, what is written never dies* is known in several languages. However, the possibility of sound recording from 1888 onwards may question the truth of the human experience summarized in that proverb. A relatively short time – about 100 years – elapsed between Edison's invention and the development of the technical background that now provides a possibility to record and save huge speech materials for phonetic and speech technology research.

Language and language use continually change due to the fact that languages live as long as the communities that speak them remain in existence. The quality and temporal course of changes also differ across periods since they are determined by social and other human factors [e.g. 1, 2]. The recognition of ongoing changes is no easy task; quite often, it is difficult to tell what is and what is not an instance of language change [3, 4]. The further away we are from a period, however, the more exactly it can be determined what qualifies as a change as compared to the present situation, what tendencies can be identified, and what possible causes may have been instrumental in it. In the case of spoken language, this is possible only if we have access to recorded speech materials that can be analyzed in multiple ways and hence compared to facts of the present state of the language.

Speech materials that were recorded for a particular purpose are immensely important. The development of large databases is due to the evolution of technology: to the emergence of computers with a large memory capacity, and the related demand for large amounts of texts recorded with specific aims in mind and encoded in a uniform manner so that researchers can analyze them in various respects. Speech databases in the current sense go back to a few decades only, both internationally [e.g. 5, 6, 7, 8] and in Hungary [e.g. 9, 10]. However, present-day corpora and databases have forerunners that are worth knowing about. We can learn from the mistakes that early developers committed as well as from their advantages and actual achievements, from their successes and failed attempts, from the knowledge they have amassed.

The first fragmental Hungarian dialect recordings we know of are found in the folk song recordings made by Béla Bartók and Zoltán Kodály, the famous Hungarian composers. In 1912, the Benedictine monk Ányos Bíró recorded dialectal speech in various locations in

Hungary specifically with the aim of handing Hungarian dialects down to posterity [11]. Probably the earliest database-like collection of spoken language recordings can be linked to the name of the phonetician József Balassa (1864–1945). The recordings were made during the First World War, under the auspices of the Phonetic Laboratory of the Academy of Eastern Trade led by Balassa. In a detention centre near Esztergom, the speech of Votyak, Bashkir, Tartar and Russian prisoners of war was recorded. The total number of subjects was 48. We know that the recordings were made to a wax cylinder by a portable phonograph. However, their cataloguing and analysis were not feasible during the war, and we do not know anything concerning their whereabouts; probably they were destroyed [12]. Yet the mere fact that such a project was carried out suggests that our ancestors felt the necessity of large-scale recording of speech.

Some three decades later, another Hungarian phonetician Lajos Hegedűs started recording dialect speech: this gave rise to the earliest extant collection in this country of speech recordings involving many speakers, known as the Hegedűs Archives today.

2. The Hegedűs Archives

2.1. The founder

It was in the nineteen-forties that, at the initiation of Lajos Hegedűs, regular production of dialect sound recordings began. Lajos Hegedűs was born on 1 March, 1908, and died young on 23 July, 1958. He was a phonetician and linguist, a disciple of Zoltán Gombocz at Budapest University, and he also studied in Vienna and in London. In 1943, he became an unpaid professor at Debrecen University; from 1944 to 1950 he worked for the Transdanubian Research Institute in Pécs (South-Western Hungary). In 1950 he was employed by the Research Institute for Linguistics of the Hungarian Academy of Sciences, and created "the museum of Hungarian dialects" with his dialect recordings. He was one of the most eminent early representatives of experimental phonetics in Hungary, and his papers are outstanding in an international perspective, too. It cannot be an accident that it was him who insisted on the importance of the creation of a collection of dialect speech recordings.

The aim of the project was to record speech materials at various locations within Hungary and in surrounding countries, in a variety of genres: narratives, storytelling, spells, etc. The aim was not a systematic collection of diverse speech styles: the main point was to record and preserve the various local dialects of Hungarian.

It is hard to believe today that Hegedűs had to insist that the significance of sound recordings, as opposed to data recorded on paper, lies in the fact that they also include suprasegmental features of speech. Let us quote one of his arguments from 60 years ago: "The material collected on gramophone records can be transposed to a kymograph by the help of an electro-magnetic apparatus and it can be submitted, in whole or part by part, to objective study concerning duration, intonation, pauses, or rhythm..." [13, p. 6]. Thus, Hegedűs was guided by three aims: to preserve samples of spoken dialects, to provide sound materials for experimental phonetic study, and (especially) to create the possibility of analyzing prosodic features of speech. Figure 1 shows the kymograph.

The first recording was made by Lajos Hegedűs in August 1940 in Drávacsehi (Baranya County). The speaker was András Száva, a 78-year-old man, reciting a tale of the Willowy Princess (8 min 22 sec). Initially, Hegedűs' work was assisted by the linguist Mihály Temesi (55 recordings), then he made recordings on his own between 1947 and 1954. Later on, other linguists joined in: József Végh, Lajos Lőrincze and Samu Imre.

Figure 1: The kymograph

The latter two also used the questionnaire of a dialect atlas then under preparation, and made sound recordings of the answers of the subjects. 58 recordings were made by the linguists Kálmán Keresztes and Lajos Balogh; in the recording of one subject each, the phonetician Iván Fónagy (later professor at Sorbonne) and the linguist György Szépe also participated. However, the name Hegedűs Archives was not randomly given, decades later, to the collection: Lajos Hegedűs personally made a total of 1538 recordings (92% of the whole material).

2.2. The subjects

The total number of subjects was 686. Due to reasons that remain unknown to us, the age of some of the subjects was not recorded, or else this piece of data did not survive in all cases. Typically, whole families were interviewed simultaneously, in some cases three generations' speech is found in a single recording. Most subjects were adults, and many were over 40 (Table 1). The oldest speaker was an 88-year-old woman; the youngest was a little girl of 10. The number of subjects under 18 years of age was 16.

Table 1: Subjects' ages and genders in the Hegedűs Archives.

Age range (years)	Number of subjects	
	females	males
71 – 88	46	36
61 – 70	30	20
51 – 60	40	28
41 – 50	52	40
31 – 40	34	12
19 – 30	21	20
16 – 18	6	5
10 – 15	12	12
Total	241	173

The number of speakers whose age is known is 414; we do not know the ages of over 150 subjects, and in a number of cases even the name of the speaker is missing, labels like "female voice", "male voice", or "unknown" were applied. We also find labels saying "several speakers": this covers conversations and/or simultaneous communication but the names and ages of the participants are missing (e.g. women chatting while spinning together). Women figure far more prominently than men do; the number of women/girls whose ages are known is 241, that of men is 173. Talkativeness, it appears, was a characteristic of women; but another reason for this asymmetry may have been that the men were doing jobs that hindered them in participation more extensively. Female subjects normally worked at home or else they worked as day labourers. The men were mainly agricultural workers, but blacksmiths, carpenters, herdsmen, etc. were also found among them.

2.3. The recordings

The total number of recordings is over 1600. They were made in a number of counties within Hungary and in areas outside the country where native Hungarians lived, in Czechoslovakia, Romania, the Soviet Union, or Austria. The fieldwork sites were in the following counties (with the number of recordings made there in parentheses): Baranya (536), Hajdú-Bihar (84), Vas (54), Nógrád (17), Tolna (394), Pest (70), Fejér (7), Zala (176), Veszprém (77), Szabolcs-Szatmár (101), Bács-Kiskun (11), Borsod-Abaúj-Zemplén (1), Maros-Torda (1), Heves (23), Nyitra (38), Abaúj-Torna (36), Gömör and Kishont (7), Ung (6), Krassó-Szörény (2), Kolozs (2), Csík (1), Somogy (3), Csongrád (19), Burgenland (1). The counties are not represented equally, and the numbers of sites in each differ widely. In some counties, 10 or more villages served as collection sites, while in some others only one or two did. The highest number of recordings comes from Baranya County (South-Western Hungary). Very few recordings were made e.g. in Fejér County (roughly the middle of the country, west of the Danube). All recordings made in Csongrád County (South-Eastern Hungary) come from a single village, Tápé. In Somogy County (west of the Danube, situated centrally) recordings were only made in Somogyhatvan, in Csík County (today in Romania), only in Csíkmadaras, and in Pest County (around the capital) only in Páty.

Recordings were dated: usually just the year was given, less frequently the month as well, and in several cases the date was given exactly, including the day when the recording was made. The first recording was made, as already mentioned, in August 1940; the last one was made on 3 May, 1957. Grey (later, pink) cards of size A/4 (14.8 × 21 cm) were made to accompany the records, containing many important pieces of information. Along with the name of the collector, these cards also revealed who produced the lacquered records themselves.

The Hegedűs Archives include a total of over 50 hours of sound recordings (approximately 600,000 running words). The duration of the shortest recording is 2 minutes, and that of the longest is 14 minutes. The topics are varied. In terms of spontaneity vs. interpretation (of pre-existing texts), they can be classified into three broad categories.

(i) Narratives: the speaker talks about an event related to his/her life, family, or job. These topics are usually raised by the field worker, and the subject talks about them spontaneously, without any previous preparation. For instance: folk customs, superstitions, festive occasions, stories from the life of the village, ways of baking bread or cakes, description of a pig-killing feast, stories on military recruitment, life stories, everyday jobs (fishing, sowing, harvest, shirt making, tin moulding, cheese making, sheep raising, etc.), wedding feasts, saint's-day fairs, healing, soldiers' stories, dispelling rats, thefts, death cases, kid's rigs, maypole raising.

(ii) Texts of varying length handed down from generation to generation in a fixed form like spells, wedding rhymes, Easter toasts, laments, and nursery rhymes. (Occasionally, the recordings also include sung sequences.)

(iii) Semi-spontaneous (or: semi-interpretative) texts where the speaker relates a story that is more or less fixed in terms of both content and linguistic form. For instance: tales, ghost stories, jokes, stories about scamps.

Narratives are sometimes interspersed with dialogues, mainly those between the subject and the field worker, but also real conversations in the sense that members of the family or neighbours who are present at the recording may interrupt the conversation every now and then.

2.4. The technology of recording

Recording involved various tools (microphones, amplifiers, power supply, record cutter, raw gramophone records, headphones, etc.); and the sessions typically required a room-sized space (Fig. 2). In the beginning, glass and aluminium-based, decilith-covered or lacquered records were used (Fig. 3).

Figure 2: Typical circumstances of speech recording in the mid-twentieth century

Recording was made such that the text spoken into the microphone was directly transmitted to the record-cutting machine, and the record was immediately ready to be played. The jackets of the individual records (Fig. 4) and (paper labels of) the records themselves had the most important data written on them: the name of the field worker, the place and time of recording, the name of the subject, the topic, and the technical data of the record produced. Some jackets exhibit Lajos Hegedűs' own handwriting. About 1700 separate recordings were made by Hegedűs and his colleagues, and they are contained in 842 original records.

Figure 3: An original lacquered record and a turntable apparatus able to play it (still in working order at the time of writing)

The quality of recordings varies. Most of them are of good quality, very moderately noisy; they are perfectly fit for instrumental analysis (see Fig. 5). In some cases, the recording is rather noisier, due to various reasons, or contains external noises (like the barking of dogs). Some of the latter are still directly analyzable, but there are others that can only be analyzed acoustically after they are stripped of background noise. But even these noisier recordings are clearly comprehensible, hence perfectly fit for linguistic or folkloristic study.

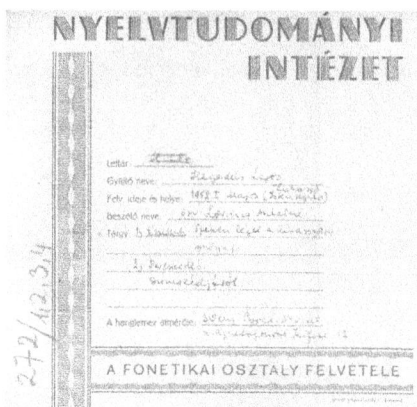

RESEARCH INSTITUTE FOR
LINGUISTICS
Inventory:
Name of collector: Hegedűs Lajos
Date and place of recording: 1952. I.
Majos
Name of subject: Lőrincz Antalné,
widowed
Topic: 1) songs (title of song)
       2) speaking about her neighbour

Diameter of record: 39 cm

RECORDING OF THE PHONETICS
DEPARTMENT

Figure 4: The jacket of the record with data on content and inventory particulars

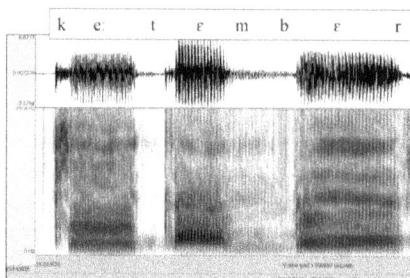

Figure 5: Oscillogram and spectrogram of the phrase *két ember* 'two people' (Praat), male speaker
(Adorjás, Baranya County)

2.5. Archiving

Most of the records were subsequently stored in the Research Institute for Linguistics of the Hungarian Academy of Sciences. Some others recently emerged as they had been in the personal possession of various linguists. Over the decades, the records started to be mildewed, and the number of record players that could play them has diminished considerably by now. The preservation of the material of the Hegedűs Archives in contemporary data carriers became rather urgent by the beginning of the twenty-first century in more respects than one.

The process of archiving consisted in two parts. 1. Technical preservation of the records, meaning both that they are physically restored and that their content is saved in contemporary storage devices in order to be playable. 2. Cataloguing the sound material, and compiling its relevant metadata in an electronically searchable database.

1. A grant from the Ministry of Culture (2003–2006) made it possible to restore and save 842 records. The restoration was made difficult by the fact that special expertise was required for cleaning operations so that neither the records nor the recordings should suffer any harm. First, the records were freed of mildew and any other sediment with a special chemical solution, then they were played by an appropriate record player and their content was recorded in a computer file. After that, the new recordings were copied, in audio format, onto CDs and

other external storage devices. The cross-compliance of the CDs and the original records was taken care of, and a system of labelling was devised for the CDs. The whole archival material is now found on the storage server of the Research Institute for Linguistics of the Hungarian Academy of Sciences, as well as on 122 CDs and two external hard discs.

2. At the same time, the metadata of the recordings were included in computer files on the basis of the notes that accompanied the original records (using the Access program). This searchable information database contains the following: serial number of the CD containing the restored sound material, track number of the CD containing the recording in question, serial number of the original record, date and place of the sound recording session (both the name of the village and that of the county), name of the subject (if available), age of the subject (if available). Furthermore, the database includes the duration of the recording (in seconds), the name of the person who made the recording, the actual topic(s) of the recording, and its definition in terms of a pre-established list of speech topics. The last-mentioned piece of information concerns the exact contents of the given recording. The topic labels are as follows: narrative, part of tale, superstition, conversation, rhymed tag, (nursery) rhyme, toast, soldier's story, dirge, dialect words (questions and answers of the dialect atlas in preparation at the time of recording), story, recital, song, and poem. These categories make it possible to search the material in terms of speech topics.

3. Studies carried out so far on the material of the Hegedűs Archives

The first portions of the Hegedűs Archives to be transcribed and analyzed by folklorists were some recordings of folk tales. The phonetic study of the recordings started in 2005 and goes on ever since. Linguistic studies concerned disfluency phenomena, prosodic features of spontaneous speech, and a comparison of various speech genres in terms of their phonetic characteristics.

Types and proportions of occurrence of disfluency phenomena were studied in recordings of 28 speakers (14 female and 14 male ones, aged between 25 and 80). The data were compared to instances of disfluency in present-day spontaneous speech by standard speakers of the same ages and genders [14]. In selecting the latter subjects, care was taken that their speech rate should be similar to that of the archival speakers. The analyses were based on a total of 2.72 hours of archival and 2.68 hours of present-day spontaneous speech material. The disfluency phenomena (of the "speaker's uncertainty" type) studied were as follows: filled pause, repetition, filler word, prolongation, pause in the word. The error types were: wrong word, contamination, restart, anticipation, perseveration, false start, grammatical error, tip of the tongue phenomenon, slip of the tongue.

The aim of the comparative study was to see if there were any differences between speakers of sixty-odd years ago and today's speakers with respect to the planning and execution processes of spontaneous speech. It was hypothesized that there would be dissimilarities in the frequency of occurrence of the individual disfluency phenomena, witnessing differences in speech planning processes, hence interpretable as signs of linguistic change. The archival material contained a total of 446 uncertainty-type disfluency phenomena, whereas that of the present-day speakers contained 1346 comparable instances. The figures include the number of silent pauses, too. The number of errors also increased in today's speakers as compared to those of the archival material: the latter exhibited 122 whereas the former exhibited a total of 408 instances. The most spectacular change occurred in the occurrence of filled pauses. They hardly ever occurred in the speech of the archival speakers, a mere 18 filled pauses were found in the whole material studied. Present-day speakers produced 636 filled pauses in their spontaneous speech of the same length. One possible reason might be that spontaneous speech is characterized by longer and more complex utterances today than it was in the fifties, as the recordings suggest. More complex constructions put a heavier load on the planning processes, and this may underlie the increase of the number of filled pauses. Another possible explanation is that the archival speakers produced far more filler words than today's speakers did, especially favoring the phrase *azt mondja* 'it says', rather than uttering hesitation noises (filled pauses). Wherever the subjects were uncertain as to what to say next, or there occurred a temporary disharmony in their speech planning processes, they most often uttered some filler words. The proportions of filler words in today's speakers are similar to those in the archival speakers (273 and 296 occurrences, respectively). A smaller but still mathematically significant difference was found in the case of repetitions (42 in the archival material, and 207 in the present-day material). Similarly, a significant difference was found in the occurrence of prolongations (84 and 140, respectively).

Archival speakers committed a total of 130 errors in their 2.7 hours of spontaneous speech, whereas present-day speakers produced 390 disfluency phenomena of the error type in the same amount of time. The number of errors attested in the present-day corpus is exactly three times as many as found with speakers of the forties or fifties. Statistical analyses confirmed that there were significant differences between the two corpora in four error types: restarts, anticipations, grammatical errors, and slips of the tongue. The discrepancy is conspicuously large in grammatical errors: 12 were attested in the archival material and 111 in the new one. The number of anticipations and slips of the tongue are seven times as large as they used to be, modified restarts and replacements were multiplied by three, and wrong words and simple restarts increased to twice the number they had in the old material.

In the archival corpus, the speakers produced 0.74 errors and 2.72 uncertainties per minute, a total of 3.46 attested cases of disfluency in a minute. Present-day subjects produced 2.53 errors and 8.17 uncertainties, a total of 8.17 attested instances of disfluency each minute. In view of the fact that the speech rates of the two populations were nearly identical, such differences in speech planning and execution give food for thought. In spite of the fact that significant differences were not found for each and every disfluency phenomenon, it is still noteworthy that the number of occurrences was larger in present-day speech in all categories, even if to different extents. The spontaneous speech of present-day speakers, then, is definitely far more spasmodic than it used to be.

Analyzing the proportions of disfluency phenomena corpus-internally, it was found that while earlier on the strategies of resolving disharmony primarily involved filler words and prolongations (in 80% of the cases), today the main exponents of such strategies are filled pauses and repetitions (in 84.6% of the cases). It appears that all this shows a change of habits in language use. With respect to errors, archival speakers exhibited more wrong words, false starts and instances of the tip of the tongue phenomenon, while present-day speakers exhibited grammatical errors, slips of the tongue, and anticipations more often than any other error types.

In present-day speakers, the relatively large number of grammatical errors, the frequent occurrence of anticipations, and a strong increase in the number of filled pauses show that they have difficulties in finding an adequate linguistic form for the expression of their thoughts. On the other hand, the speech planning processes of the archival speakers proved to be more hampered by difficulties in activating their mental lexicon than in creating the appropriate grammatical form for what they wanted to say.

Potential changes in articulation rate and speech rate were explored on the basis of the Hegedűs Archives, the BEA database [10] containing contemporary standard speech samples, and recordings of subjects from a village in Nógrád County [15]. The results confirmed that there are significant differences between the archival speakers and both present-day populations (village-dwellers and inhabitants of the capital) in articulation rate (mean values: 9.9 sounds/s, 11.1 sounds/s, and 12.6 sounds/s, respectively). In speech rate, there was no difference between archival speakers and today's village people, but both groups differed significantly from the Budapest speakers (7.9 sounds/s, 8 sounds/s, and 9.7 sounds/s). The comparison showed that there were differences in pause durations across all three groups. Silent pauses were the longest in recordings of the Hegedűs Archives (mean: 653 ms), shorter than that in the speech of villagers (mean: 596 ms), and the shortest in that of present-day Budapest speakers (mean: 518 ms). Archival speakers paused primarily at phrase boundaries (in 65% of the cases), as opposed to today's speakers who only paused at those boundaries in 52% of all cases. The study appears to confirm a relative acceleration of speech rate over the decades, at least with respect to those groups.

Temporal properties of the speech of fourteen young subjects of the Archives aged 10 to 12 and 14 to 16, respectively, in spontaneous speech versus storytelling [16]. The results show that storytelling is slower than spontaneous narratives are, and that the speech rate of older children is slightly higher than that of younger children. Pauses occur more often in narratives than in storytelling. Both articulation rate and speech rate are somewhat faster in these children than in adult speakers of the Archives.

Recordings of the Hegedűs Archives also give us an excellent chance for studying the various genres of speech (tales of various sorts, spells, superstitions, descriptions of folk customs, spontaneous narratives, etc.). Menyhárt [17]'s investigations showed that the speakers' speech rate was the fastest in relating fairy tales, and the slowest in spontaneous narratives. Pitch range, too, was the widest in fairy tales. The lowest fundamental frequency was found in spontaneous speech, and the highest in tales about animals. Archival speakers produced the briefest pauses in ritual(ized) texts; the average durations of silent pauses were nearly equal in tales and in spontaneous narratives.

4. Outlook

In the second half of the twentieth century, large speech databases were created in a number of languages and for a variety of purposes. However, there are few if any corpora that are instrumentally analysable but contain material coming from the mid-twentieth century and that have been created by linguists primarily for the purposes of phonetic studies. The main significance of the Hegedűs Archives lies in the fact that it makes it possible for us to study the properties of spontaneous and semi-spontaneous speech of over 70 years ago, to compare them with present-day speech, and that the recordings can also be studied in sociolinguistic, dialectological, and ethnographic terms. The contents of the speech materials are part of our cultural heritage. The Hegedűs Archives can be taken to be an up-to-date database as it consists of a searchable written database and an organised speech material, making all sorts of utilisations easy to perform. Full annotation of the whole material of the Archives, however, remains a task for the future.

References

[1]   Wardhaugh, R.: An introduction to sociolinguistics. Malden MA: Blackwell, 2006.

[2]   Ohala, J. J.: The listener as a source of sound change (perception, production, and social factors). In Solé, Maria-Josep and Recasens, Daniel (eds.): The initiation of sound change. Amsterdam: John Benjamins, 2012, 21–36.

[3]   Pettigrew, A. M.: Longitudinal field research on change: Theory and practice. Organization Science 1 (1990), 267–292.

[4]   Bybee, J.: Formal universals as emergent phenomena: The origins of structure preservation. In Good, Jeff (ed.): Linguistic universals and language change. Oxford: Oxford University Press, 2008, 108–121.

[5]   Garofolo, J.; Lamel, L.; Fisher, W.; Fiscus, J.; Pallet, D.; Dahlgren, N.: Darpa, TIMIT, Acoustic-phonetic continuous speech corpus. Washington DC: US Department of Commerce, 1993.

[6]   Simpson, A.; Kohler, K. J.; Rettstadt, T. (eds.): The Kiel Corpus of read/spontaneous speech – acoustic database, processing tools and analysis results. Arbeitsberichte des Instituts für Phonetik und digitale Sprachverarbeitung der Universität Kiel (AIPUK), 32, 1997.

[7]   Burnard, L.; Aston, G.: The BNC handbook: exploring the British National Corpus. Edinburgh: Edinburgh University Press, 1998.

[8]   Grønnum, N.: A Danish phonetically annotated spontaneous speech corpus (DanPASS). Speech Communication 51 (2009), 594–603.

[9]   Váradi, T. 2003. A Budapesti Szociolingvisztikai Interjú. In Kiefer, F.; Siptár, P. (eds.): A magyar nyelv kézikönyve [A handbook of the Hungarian language]. Budapest: Akadémiai Kiadó, 339–359.

[10]   Gósy, M.: BEA – A multifunctional Hungarian spoken language database. The Phonetician 105/106 (2012), 50–61.

[11]   Kiss, J. ed.: Magyar dialektológia [Hungarian dialectology]. Budapest: Osiris Kiadó, 2001.

[12]   KKA: Ismertetés a Keleti Kereskedelmi Akadémia Fonetikai Laboratóriumának munkájáról [A report on the work of the Phonetics Laboratory of the Academy of Eastern Trade]. In: A KKA 25. évi jelentése az 1915-16-iki iskolaév végén. Budapest, 1916. 55–56. Reprinted in: Studia Academiae Nyíregyháziensis, Tomus III, 1994.

[13]   Hegedűs, L.: Népi beszélgetések az Ormánságból [Conversations of village dwellers from South-West Hungary]. Pécs: Szabadság Pécsi Nyomda és Könyvkiadó Kft, 1946.

[14]   Gósy, M.; Gyarmathy, D.: A nyelvhasználati változás egy jelensége [An instance of change in language use]. Magyar Nyelvőr 132 (2008), 206–222.

[15]   Menyhárt, K.: A spontán beszéd változásai időben és térben [Spatial and temporal variability in spontaneous speech]. In Borbély, A.; V. Kremmer, I.; Hattyár, H. (eds.): Nyelvideológiák, attitűdök és sztereotípiák. 15. Élőnyelvi Konferencia [Linguistic ideologies, attitudes, and stereotypes. Papers from the 15th Conference on Spoken Language]. Budapest: Tinta Kiadó, 2009, 111–119.

[16]   Menyhárt, K. 2012. A beszéd temporális jellemzői 60 évvel ezelőtti gyermek beszélőknél [Temporal characteristics of children's speech 60 years ago]. Beszédkutatás 2012. 246–259.

[17]   Menyhárt, K. 2010. Folklór és fonetika. Mesék, ráolvasások és más szövegek akusztikai-fonetikai vizsgálata [Folk lore and phonetics: An acoustic phonetic investigation of tales, spells and other texts]. In Szemerkényi, Ágnes (ed.): Folklór és nyelv [Folk lore and language]. Budapest: MTA Néprajzi Kutatóintézet, 27–40.

The Early Swiss Dialect Recording Collection "LA" (1924–1927): A Description and a Work Plan for Its Comprehensive Edition

D. Studer-Joho

Universität Zürich, Institut für Vergleichende Sprachwissenschaft: Phonogrammarchiv
dieter.studer@access.uzh.ch

Abstract: Between 1924 and 1927 – and again in 1929 – the Phonogram Archives of the University of Zurich collaborated with Prof. Dr. Wilhelm Doegen of the *Lautabteilung der Preussischen Staatsbibliothek zu Berlin* in a dialectological recording campaign. Each year, Doegen was invited to travel to Switzerland from Berlin for a few days with his heavy recording equipment to collect a few dozen dialect specimens recorded with speakers from all over Switzerland. Before each recording session suitable speakers were carefully selected – mainly based on dialectological criteria – and were asked to prepare a short text for recitation (usually in their vernacular), which they then delivered into Doegen's phonograph. All in all, some 225 shellac records could be produced in this fashion, before Doegen fell in disfavour in Berlin and could no longer continue his collaboration with Zurich. While the recordings from the later so-called "LM" campaign in 1929, which were collected in the southern Swiss town of Bellinzona and the northern Italian town of Domodossola, have either already been published or are in the process of being published, the ca. 175 recordings from the 1924–1927 "LA" campaign, collected in Zurich, Bern, Chur, Sion and Brig, respectively, have been published only partially. This paper discusses the possibilities in making this precious Swiss dialectological treasure trove available both to the interested public and to dialectological research.

1. Introduction

The Phonogram Archives of the University of Zurich – the oldest audio-visual archive of Switzerland – started collecting Swiss dialect recordings with a phonograph apparatus acquired from the Vienna Phonogram Archives in 1909. The close collaboration with Vienna lasted from then until 1923, and some 318 wax phonograms could be collected and preserved for posterity during these years. The lasting fruit of this successful collaboration were comprehensively edited in 2002 [1]. The Vienna wax phonograph – albeit an ingenious lightweight and reliable contraption – could no longer be considered state-of-the-art in the 1920s with respect to sound quality, so the Zurich Phonogram Archives decided to look for a new technological partner and invited Prof. Dr. Wilhelm Doegen (1877–1967) of the *Lautabteilung der Preussischen Staatsbibliothek zu Berlin* to conduct recording sessions with his gramophone recorder in Zurich during five consecutive days in June 1924. The results were promising and similar recording sessions under Doegen's supervision were conducted in Bern in 1925, in Chur in 1926 and in Sion and Brig in 1927. During these four campaigns some 175 recordings were collected and they were shelfmarked "LA" [="Lautarchiv"]. The following year, Doegen fell ill and the planned recording sessions in Bellinzona and Domodossola (I) had to be deferred to 1929, during which another 54 recordings could be made. This second recording campaign in the Italian-speaking parts of Switzerland and the German-speaking linguistic enclaves of North-Western Italy, collectively referred to as "LM" [="Lombardische Mundarten"] turned out to mark the end of the collaboration with Berlin, as Doegen fell in disfavour in Berlin and was suspended – first temporarily in 1930 and later permanently in 1933 (cf. [2]). The idea behind the collection of the material was of course the publication – both as sound and text – of linguistic material of dialectological interest for subsequent research. The 1929 "LM" recording campaign comprised both a number of Highest Alemannic German recordings, which were published along with phonetic transcriptions and a linguistic commentary in [3], and a group of Ticinese vernacular (Lombard Italian) dialect recordings, which are currently being prepared for publication as [4]. More than eighty years after its collection, however, most of the material of the "LA" campaign (1924–1927) is not readily available for research.

2. Characterization of the Swiss "LA" Collection

In preparation of the actual recording sessions, potentially suitable speakers were contacted with the request to prepare a short text for the recording session:

> For speech recordings a vernacular text – if possible composed by yourself – of about three minutes' duration would be suitable. It might be about a local legend, some event, some incident or the like; for a musical recording perhaps a traditional folk song with vernacular lyrics, or perhaps a popular instrumental piece of music. It goes without saying that it is our endeavour to get genuine, vernacular dialect in speech recordings and genuinely popular tunes in music recordings, exclusively. [5][1]

Accordingly, the majority of recordings consist of a prepared monologue, in which the speaker relates a short narrative. Since the requirements (i.e. preparing a short text for reading) were fairly demanding, it is no surprise that teachers and writers are somewhat overrepresented in the sample. Also, the majority of speakers are elderly males, but younger males and women are also included invariably, so that the social characteristics can neither be said to be completely uniform nor completely random. Not all speakers coped equally well with the difficult situation of reading out loud a text they had written down in the vernacular – a rather artificial situation for most speakers –, so that some of the recordings contain a fair share of stutters and interruptions. Additional pressure must surely have been created by the fact that there was no possibility of editing the recording; the first take had to be perfect. Only in a few instances the recording was repeated; presumably the first recording was deemed unacceptably bad in these cases. Overall, the artificial situation of the recording resulted in a rather elevated – perhaps even literary – style of dialect and only in some acted-out dialogues more informal registers can be glimpsed. Compared to the Vienna wax discs, the quality of the recordings is generally quite good; some recordings feature a characteristic repetitive noise connected to the turning of the disc, but in general the intelligibility is quite good.

3. The Original Shellac Discs and Their Extant Documentation

Luckily most of the Swiss "LA" discs are still extant and in very good shape; LA 261[2] was accidentally destroyed during the production of the metal matrix, but the same speaker could

---

[1] Author's translation; the original reads: "Für Sprechaufnahmen käme ein womöglich von Ihnen selbst verfasster Mundarttext von etwa 3 Minuten Sprechdauer in Frage, der eine örtliche Sage, Begebenheit, ein Erlebnis odgl. zum Inhalt hätte, für Musikaufnahmen ein Volkslied mit mundartlichem Text, etwa auch ein volkstümliches Instrumental-Stück. Dass unser Bestreben dahin geht, bei Sprechaufnahmen nur echte, bodenständige Mundart, bei Musikaufnahmen nur wirklich Volkstümliches, zu erhalten, braucht kaum hinzugefügt werden."

[2] The records collected by the *Lautabteilung* were numbered consecutively; the shelfmarks of the Swiss "LA" collection are as follows: 1924 (Zürich) records run from "LA 250" to "LA 281"; 1925 (Bern) from "LA 521" to "LA 561"; 1926 (Chur) from "LA 772" to "LA 822"; and 1927 (Sion and Brig) from "LA 911" to "LA 960" (cf. also the appendix).

repeat his contribution in 1927 as LA 959. Although, the Phonogram Archives do not hold a complete set of the collection, a complete set is available in the original *Lautarchiv* in Berlin (now part of the collections of the Humboldt University). Thanks to the indefatigable efforts of Jürgen Mahrenholz, who supervised the digitization of the Lautarchiv around 2004, the Phonogram Archives have digital copies of all its extant and missing "LA" recordings. The digital files are of good quality and usually yield satisfying results after minor sound restoration procedures (declick, dehum, decrackle). Back in 2004 the discs were digitized at 44.1kHz/16Bits, which provides some headroom for post-production, but for archival purposes much higher sampling rates would now be applied. The limiting factor, however, is after all the shellac disc, which, although a reliable recording medium, inherently comes with certain shortcomings that no restauration procedure can undo. These shortcomings are thus part of the recording and with a little acclimatization the human ear is capable of ignoring a good deal of the noise that is present. It is also striking to observe that the availability of transcriptions often greatly enhances the intelligibility of the recordings.

Luckily, the Phonogram Archives are also in the possession of documentation that was produced along with the actual recordings, such as correspondence with and personal records of the subjects, text drafts and phonetic transcriptions of the delivered texts and a number of written reports that provide a context in which the recordings themselves can be more easily understood (not only acoustically). Translations are also often available and can provide very welcome hints as to the exact meaning of individual words or phrases that are difficult to hear. Currently we are labelling and scanning all the documentation, both in order to have a backup (in case of a fire) but also to get a guide as to how the many slips of paper, which often have differing formats, originally were positioned next to each other in the various folders and dossiers.

4. Previous Partial Publication of the Material

It is interesting to see that the publication of the recorded material is clearly biased in favour of the Romance-speaking parts of the country. Possibly, this is due to the fact that the vernaculars in these areas are more under pressure from either the respective standard language (French vs. Franco-Provençal; standard Italian vs. vernacular Ticinese Italian) or from Swiss German, in the case of Romansh; thus, the dialectological interest in recorded speech is perhaps more acute than in the German-speaking part of Switzerland where the use of the vernacular Swiss German dialects is firmly embedded in the majority of communicational contexts.

The 1924 recordings from Zurich were only marginally incorporated in Doegen's series *Lautbibliothek* [6]; the remainder of that campaign remain unpublished. About half of the 1925 recordings from Bern, however, were edited with transcriptions in the "Lautbibilothek". The 1926 campaign in Chur is the best-documented campaign, as the majority of the recordings from it were published in [7], [8] and [9], respectively (the latter two also including digital files of the sound). The Franco-Provençal recordings from Valais 1927 were edited almost exhaustively in [10]; of the German-speaking recordings of the same year only a selection of four recordings were edited in [11]. Thus, only about half of the recordings are available in printed editions in various places and formats, and less than a fourth of all recordings are currently available as digital sound files.

5. Plans for the Comprehensive Edition of the Unpublished Material

It is the Phonogram Archives' firm resolution to make all Swiss "LA" recordings accessible to the interested public in general and to dialectological research in particular in the next few years. However, experience with recent edition projects has shown that it is very important to divide the collection into manageable parts, because the edition process puts a considerable strain on the small archive with its two 50% part-time staff, and editorial problems seem to grow exponentially with the number of recordings tackled at once. Moreover, the practice of producing a printed book with added audio-CDs has also been put into question lately, as more modern manners of (e-)publication may in fact reduce some of the financial strain on the archive. Digital publication may also allow for different formats of the transcriptions (e.g. phonetic TextGrids or TEI/XML corpus formats) in addition to a printable output, which may prove interesting in view of potential computational applications at some later stage.

Doegen's *Lautbibliothek* may offer an interesting model for a productive publication of the material: each recording is presented in a fixed format including orthographic transcription, phonetic transcription and translation on a folded paper in landscape orientation. These booklets could then be served for download along with the digital file of the recording. Printing such booklets is also very cost-efficient and after a sufficient number of such booklets have been created, they could be bound together as fascicles into a printed volume to create a tangible residue from the virtual booklets in due time. The booklet approach would also ensure that there would be some output even if the task of publishing all of the material should prove to be too daunting a task after all. Work on the material could be interrupted and continued at some later stage again.

6. Appendix: Overview of recordings

The manner of the following presentation has a dialectological bias, as it only lists the languages spoken and the geographic locality associated with the variety spoken in the recording (followed by the abbreviated canton in brackets), as well as the names of the speakers. Some recordings contain yodelling or other non-verbal forms of music, or the dialect presented in the song cannot be said to be associated with any locality as such, but represents a mesolectal form of the language; in such cases the locality is enclosed in square brackets. The right-most column lists previous publications of the respective recording.

Abbreviations used
bar: Bavarian German.   *BWe*: Gadmer (2012) [8].   deu: Standard German.
Fef: Valär (2013) [9].   frp: Franco-Provençal.   fra: (vernacular) langue d'oïl French.
gsw: Swiss German.   lmb: Ticinese dialect.   *LB*: Doegen (1929–1938) [6] (no audio).
non-v.: non-verbal.   roh: Romansh.   *RurM*: Schorta (1946) [7] (no audio).
SdM: Dieth (1951) [11].

A1) Zurich (ZH), 13–17 June 1924, premises of the Phonogram Archives of the University of Zurich.

LA 250	gsw	Einsiedeln (SZ)	Meinrad Lienert	
LA 251	gsw	Zürich (ZH)	Paul Usteri	
LA 252	gsw	Zürich (ZH)	Henri Mousson	*LB*: 101.
LA 253	gsw	Bertschikon (ZH)	Alfred Huggenberger	
LA 254	gsw	[singing]	Hanns In Der Gand	
LA 255	gsw	[singing]	Hanns In Der Gand	
LA 256	gsw	[singing]	Hanns In Der Gand	
LA 257	frp	Montana (VS)	Baptiste Rey, Ernest Berclaz	
LA 258	gsw	Dättlikon (ZH)	Heinrich Ernst	
LA 259	gsw	Luchsingen (GL)	Jakob Hefti	
LA 260	frp	Crésuz (FR)	Cyprien Ruffieux	
LA 261	gsw	Brig-Glis (VS)	Walter Henzen	

LA 262	gsw	Göschenen (UR)	Albert Jutz	
LA 263	frp	[Neirivue (FR)]	Clement Castella	
LA 264	gsw	Bosco/Gurin (TI)	Hans Tomamichel	*LB*: 150.
LA 265	frp	Rovray (VD)	Octave Chambaz	
LA 266	fra	[Buix (JU)]	Camille Courbat	
LA 267	fra	Charmoille (JU)	François-Joseph Fridelance	
LA 268	gsw	[singing]	Piet Deutsch	
LA 269	gsw	[singing]	Piet Deutsch	
LA 270	gsw	[singing]	Piet Deutsch	
LA 271	gsw	Kirchleerau (AG)	Emil Linder	
LA 272	gsw	Zürich (ZH)	Hilde Bachmann	
LA 273	roh	Riom-Parsonz (GR)	Gion Men Collet	
LA 274	roh	Disentis/Mustér (GR)	Felix Huonder	
LA 275	gsw	Stein (AR)	Jost Küng	
LA 276	roh	Scuol (GR)	Men Rauch	
LA 277	non-v.	[Appenzell (AI)]	Emil Fritsche	
LA 278	gsw	Zürich (ZH)	Friedrich Otto Pestalozzi	
LA 279	gsw	Sevelen (SG)	Leonhard Hagmann	
LA 280	roh	Sent (GR)	Chasper Pult	
LA 281	frp	La Chaux-de-Fonds (NE)	Louis Gauchat	

A2) Bern (BE), 19–23 Sept. 1925, music room of the teacher's training college (*Ober–seminar*).

LA 521	gsw	Bern (BE)	Rudolf von Tavel	*LB*: 100.
LA 522	gsw	Langenthal (BE)	Karl Jaberg	*LB*: 105.
LA 523	gsw	Basel (BS)	Paul Speiser	
LA 524	gsw	Basel (BS)	Eberhard Vischer	*LB*: 102.
LA 525	gsw	Lenk (BE)	Hans Allemann-Wampfler	*LB*: 122.
LA 526	gsw	Grindelwald (BE)	Samuel Brawand	
LA 527	frp	Bern (BE)	Louis Gauchat	
LA 528	gsw	Diessbach bei Büren (BE)	Otto Spielmann	*LB*: 115.
LA 529	gsw	Rüttenen (SO)	Josef Reinhart	*LB*: 116.
LA 530	gsw	Schüpfheim (LU)	Franz Zihlmann	
LA 531	gsw	Saanen (BE)	Robert Marti-Wehren	*LB*: 114.
LA 532	frp	Savigny (VD)	Jules Cordey	
LA 533	gsw	Meiringen (BE)	Fritz Leuthold	
LA 534	frp	Savièse (VS)	Basile Luyet	
LA 535	frp	Savièse (VS)	Basile Luyet	
LA 536	frp	Conthey (VS)	René Jaquemet	
LA 537	frp	Granges-de-Vesin (FR)	Augustin Rey	
LA 538	gsw	Vingelz (BL)	Fritz Römer	
LA 539	gsw	Egerkingen (SO)	Eduard Fischer	
LA 540	non-v.	[Lauenen (BE)]	Katharina Hauswirth	
LA 541	non-v.	[Lauenen (BE)]	Katharina Hauswirth	
LA 542	gsw	Guggisberg (BE)	Peter Burri	
LA 543	gsw	Guggisberg (BE)	Arnold Kohli	*LB*: 104.
LA 544	gsw	Bärschwil (SO)	Albin Fringeli	*LB*: 124.
LA 545	gsw	Lauenen (BE)	Elise Perreten	
LA 546	gsw	Interlaken (BE)	Emma Spreng-Reinhardt	
LA 547	gsw	Reitnau (AG)	Reinhard Meyer	*LB*: 117.
LA 548	gsw	Guttannen (BE)	Albert Joh. Brüschweiler	*LB*: 121.
LA 549	gsw	Böningen (SO)	Hans Michel	
LA 550	gsw	Laupen (BE)	Emil Balmer	*LB*: 106.

LA 551	gsw	[Lauenen (BE)]	[mixed choir of four voices]	
LA 552	gsw	Zürich (ZH)	Oskar Wettstein	*LB*: 112.
LA 553	gsw	Frutigen (BE)	Gottlieb Trachsel	*LB*: 103.
LA 554	gsw	Lüscherz (BE)	Gottfried Grimm	
LA 555	non-v.	[Bern (BE)]	[group of musicians]	
LA 556	gsw	Muttenz (BL)	Fritz Gysin	*LB*: 119.
LA 557	gsw	Zweisimmen (BE)	Samuel Imobersteg	
LA 558 a	gsw	Reigoldswil (BL)	Gustav Schneider	*LB*: 118.
LA 558 b	gsw	Oberdorf (BL)	Paul Suter	*LB*: 118.
LA 559	gsw	Aesch (BL)	Hans Meyer	*LB*: 113.
LA 560	gsw	Sissach (BL)	Walter Schaub	*LB*: 120.
LA 561	gsw	Wenslingen (BL)	Traugott Meyer	

A3) Chur (GR), 13–18 Sept. 1926, administration building of the *RhB* (*Rhätische Bahn)* rail company.

LA 772	gsw	Tschappina (GR)	Christian Schuhmacher	*BWe*: 17.
LA 773	gsw	Küblis (GR), Seewis (GR)	Georg Hitz, Hans Brunner	*LB*: 110; *BWe*: 1.
LA 774	gsw	Peist (GR)	Nini Brunold	*BWe*: 10.
LA 775	gsw	Chur (GR)	Elise von Salis-Tscharner	
LA 776	gsw	Flums (SG)	Justus Senti	
LA 777	bar	Samnaun (GR)	Ludwig Jenal	*LB*: 123.
LA 778	gsw	Klosters-Serneus (GR)	Johannes B. Gartmann	*BWe*: 3.
LA 779	non-v.	Furna (GR)	[two singers]	*BWe*: 2.
LA 780	gsw	Safien (GR)	Leonhard Bandli	*BWe*: 16.
LA 781	non-v.	Furna (GR)	[two singers]	*BWe*: 8, 14, 18.
LA 782	gsw	Langwies (GR)	Jann Danuser	*LB*: 109; *BWe*: 9.
LA 783	gsw	Obersaxen (GR)	Martin Mirer	*BWe*: 20.
LA 784	gsw	Davos (GR)	Christian Bernhard	*LB*: 108; *BWe*: 4.
LA 785	gsw	Obersaxen (GR)	Martin Mirer	*BWe*: 20.
LA 786	gsw	Jenins (GR)	Johannes Lampert	
LA 787	gsw	Davos (GR)	Andreas Laely	*BWe*: 5, 6.
LA 788	roh	Bonaduz (GR)	Lucius Fidelis Maron	*RurM*: 1/2; *Fef*: 5.
LA 789	gsw	Arosa (GR)	Bartholomae Mettier	*BWe*: 7.
LA 790	gsw	Vals (GR)	Josef Jörger	*LB*: 111; *BWe*: 15.
LA 791	gsw	Bonaduz (GR)	Lucius Fidelis Maron	
LA 792	gsw	Praden (GR)	Nina Lyss	*BWe*: 12.
LA 793	roh	Innerferrera (GR)	Georg Mani	*RurM*: 1/3; *Fef*: 1.
LA 794	roh	Mathon (GR)	Tumasch Dolf	*RurM*: 1/4; *Fef*: 8.
LA 795	roh	Domat/Ems (GR)	Balzer Theus	*RurM*: 1/5; *Fef*: 4.
LA 796	lmb	Poschiavo (GR)	Attilio Mengotti	*RurM*: 2/1.
LA 797	roh	Alvaneu (GR)	Arthur Balzer	*RurM*: 1/8; *Fef*: 11.
LA 798	roh	Alvaneu (GR)	Arthur Balzer	*RurM*: 1/8; *Fef*: 11.
LA 799	roh	Vrin (GR)	Rest-Antoni Solèr	*LB*: 158; *Fef*: 3.
LA 800	roh	Rueras (GR)	Baschi Berther	*LB*: 154; *Fef*: 1.
LA 801	roh	Breil/Brigels (GR)	Mattias Cabialaveta	*RurM*: 1/10; *Fef*: 2.
LA 802	roh	Breil/Brigels (GR)	Mattias Cabialaveta	*RurM*: 1/10; *Fef*: 2.
LA 803	lmb	Bondo (GR)	Reto Picenoni	*RurM*: 2/2.
LA 804	gsw	Valendas (GR)	Martin Bandli	*BWe*: 19.
LA 805	roh	Sarn (GR)	Ruben Lanieca	*RurM*: 1/11; *Fef*: 6.
LA 806	gsw	Churwalden (GR)	Jakob Hemmi	*BWe*: 13.
LA 807	roh	Obervaz (GR)	Nicol Jochberg	*LB*: 157; *Fef*: 10.
LA 808	roh	Mon (GR)	Adolf Bossi	*RurM*: 1/6; *Fef*: 12.
LA 809	deu	[non-vernacular]	Gustav Bener	

LA 810	roh	Savognin (GR)	Pedar Spinatsch	*RurM*: 1/7; *Fef*: 13.
LA 811	roh	Marmorera (GR)	Emil Ghisletti	*RurM*: 1/9; *Fef*: 14.
LA 812	roh	Zernez (GR)	Andrea Schorta	*Fef*: 18.
LA 813	roh	Bergün/Bravuogn (GR)	Josti Juvalta	*RurM*: 1/1; *Fef*: 15.
LA 814	roh	Scharans (GR)	Georg Gees	*RurM*: 1/12; *Fef*: 7.
LA 815	gsw	Tamins (GR)	Ulrich Farber	
LA 816	roh	Vnà (GR)	Andrea Semadeni	*LB*: 155; *Fef*: 21.
LA 817	roh	Zuoz (GR)	Maria Schucan	*RurM*: 1/14; *Fef*: 17.
LA 818	gsw	Castiel (GR)	Sebastian Pieth	*BWe*: 11.
LA 819	roh	Scuol (GR)	Johann-Otto Ranel	*RurM*: 1/13; *Fef*: 20.
LA 820	roh	Celerina/Schlarigna (GR)	Augusta-Cecilia Pool	*RurM*: 1/15; *Fef*: 16.
LA 821	roh	Valchava (GR)	Bartholomäus Pünchera	*LB*: 156; *Fef*: 19.
LA 822	roh	[Chur (GR)]	[male choir]	*Fef*: 22.

A4a) Sion (VS), 19–21 Sept. 1927, *Musée Industriel.*

LA 911	frp	Evolène (VS)	Marie Métrailler	*LB*: 68.
LA 912	frp	Troistorrents (VS)	Adrien Martenet	*LB*: 52.
LA 913	frp	Sion (VS)	Rémy Vannoy-Planchamp	*LB*: 51.
LA 914	frp	Val-d'Illiez (VS)	Antoine Rey Hermet	*LB*: 53.
LA 915	frp	Isérables (VS)	Emile Gillioz	*LB*: 58.
LA 916	frp	Lens (VS)	François Lamon	*LB*: 69.
LA 917	frp	Grône (VS)	Albert Devantéry	*LB*: 71.
LA 918	frp	Evolène (VS)	Martin Beytrison	*LB*: 68.
LA 919	frp	Salvan (VS)	Maurice Gross	*LB*: 54.
LA 920	frp	Nendaz (VS)	Pierre Lathion	*LB*: 62.
LA 921	frp	Liddes (VS)	César Marquis	*LB*: 59.
LA 922	frp	Hérémence (VS)	Pierre Dayer	*LB*: 66.
LA 923	frp	Le Bouveret (VS)	Cyrille Curdy	*LB*: 50.
LA 924	frp	Fully (VS)	Clément Bender	*LB*: 57.
LA 925	frp	Sion (VS)	Séraphin Bétrisey	*LB*: 65.
LA 926	frp	Martigny (VS)	Jean-Pierre Moret	*LB*: 56.
LA 927	frp	Saint-Martin (VS)	Julien Mayor	*LB*: 67.
LA 928	frp	Vissoie (VS)	Remy Monnier	*LB*: 72.
LA 929	frp	Nendaz (VS)	Pierre Joseph Michelet	*LB*: 62.
LA 930	frp	Lourtier (VS)	Maurice Gabbud	
LA 931	frp	Miège (VS)	Gaspard Caloz	*LB*: 67.
LA 932	frp	[Verbier (VS)]	Léonce Gailland	*LB*: 60.
LA 933	frp	Verbier (VS)	Léonce Gailland	*LB*: 60.
LA 934	frp	Chamoson (VS)	Joseph Carruzzo	*LB*: 57.
LA 935	frp	Vérossaz (VS)	Alexis Coutaz	*LB*: 54.

A4b) Brig (VS), 22–24 Sept. 1927, Brig Town Hall.

LA 936	gsw	Münster-Geschinen (VS)	Adolf Werlen	
LA 937	gsw	Betten (VS)	Johann Mangisch	*SdM*: 1
LA 938	gsw	Zeneggen (VS)	Adolf Henzelmann	
LA 939	gsw	Salgesch (VS)	Theophil Montani	
LA 940	gsw	Leukerbad (VS)	Konstantin Grichting	
LA 941	gsw	[Brig-Glis (VS)]	[choir of four male voices]	
LA 942	gsw	Leukerbad (VS)	Alois Steiner	*SdM*: 3
LA 943	gsw	Naters (VS)	Anton Schmidt	
LA 944	gsw	Unterems (VS)	Paul Zeiler	
LA 945	gsw	Oberwald (VS)	Alex Hischier	*SdM*: 2

LA 946	gsw	Münster-Geschinen (VS)	Camil Lagger		
LA 947	gsw	Fiesch (VS)	Katharina Schmidt		
LA 948	gsw	St. Niklaus (VS)	Emil Imboden		
LA 949	gsw	Sion (VS)	Adolphe Favre	*SdM*: 4	
LA 950	gsw	Turtmann (VS)	Leo Meyer		
LA 951	gsw	Simplon (VS)	Emanuel Arnold		
LA 952	gsw	Bürchen (VS)	Alois Gattlen		
LA 953	gsw	Ausserberg (VS)	Michael Heynen		
LA 954	gsw	Saas-Grund (VS)	Peter Joseph Anthamatten		
LA 955	gsw	Grafschaft (VS)	Otto Biderbost		
LA 956	gsw	Binn (VS)	Theodor Walpen		
LA 957	gsw	Visperterminen (VS)	Heinrich Ambort		
LA 958	gsw	Wiler (Lötschen) (VS)	Otto Roth		
LA 959	gsw	Tafers (VS)	Walter Henzen	*LB*: 107.	
LA 960	gsw	[Brig-Glis (VS)]	[mixed choir of five voices]		

References

[1] Fleischer, J.; Gadmer, T.: Schweizer Aufnahmen – Enregistrements Suisses – Ricordi sonori Svizzeri – Registraziuns Svizras (deutsch, français, italiano, rumantsch) (= Tondokumente aus dem Phonogrammarchiv der Öst. Akad. der Wiss.. Gesamtausg. der Hist. Best., Series 6). OEAW PHA CD 16–18. Wien: Öst. Akad. der Wissenschaften – Zürich: Phonogrammarchiv der Univ. Zürich, 2002.

[2] Mahrenholz, J.: Zum Lautarchiv und seiner wissenschaftlichen Erschliessung durch die Datenbank IMAGO. In: Bröcker, M. (ed.): Berichte aus dem ICTM-Nationalkomitee Deutschland 12. Bamberg: Universitätsbibliothek, 2003, 131–152.

[3] Gysling, F.; Hotzenköcherle, R.: Walser Dialekte in Oberitalien in Text und Ton. Huber: Frauenfeld, 1952.

[4] Bernardasci, C.; Schwarzenbach, M.: Stòri, stralùsc e stremizzi: Registrazioni storiche (1929) della Svizzera italiana, in press.

[5] Gröger, O.: Letter to a group of teachers from the Bernese Oberland, 15. Juni 1925. Briefe Auslauf II, 1922–1929 (no. 127).

[6] Doegen, W. (ed.): Lautbibliothek: Phonetische Platten und Umschriften herausgegeben von der Lautabteilung an der preussischen Staatsbibliothek. Berlin: Preussische Staatsbibliothek, various vols. from 1929–1938.

[7] Schorta, A.: Rätoromanische und rätolombardische Mundarten. Frauenfeld: Huber, 1946.

[8] Gadmer, T. (ed.): Bündner Walser erzählen: Sprachaufnahmen aus dem Jahr 1926. 2nd ed. Zürich: Phonogrammarchiv der Universität Zürich, 2012.

[9] Valär, R. (ed.): Filistuccas e fafanoias da temp vegl – Flausen und Fabeleien aus alter Zeit: Registraziuns dialectalas rumantschas – Rätoromanische Mundartaufnahmen. Andrea Schorta | 1926. Chur: Institut dal Dicziunari Rumantsch Grischun – Zürich: Phonogrammarchiv der Univ. Zürich, 2013.

[10] Jeanjaquet, J.; Tappolet, E.: Vingt-Cinq Textes Patois du Valais: Enregistrés au Gramophone. Frauenfeld: Huber, 1929–1938.

[11] Dieth, E. (ed.): Schweizer Dialekte in Text und Ton: I. Schweizerdeutsche Mundarten, Heft 1/2. Frauenfeld: Huber, 1951. Recordings available on audio-CD (2000).

The Prague historical collection of tuning forks: A surviving replica of the Koenig tonometre

Pavel Šturm

Charles University in Prague, Institute of Phonetics
pavel.sturm@ff.cuni.cz

Abstract: Despite the copious advances in acoustic techniques and devices for measuring and displaying acoustic parameters, a steady component of a phonetician's scientific method remains to be – to this very day – our hearing. This paper therefore focuses on a historical device shared between acoustics and auditory phonetic analysis: Rudolph Koenig's Grand Tonometre, consisting of more than a hundred tuning forks, which occupies an exceptional position among the vast array of instruments manufactured by Koenig. One copy of the original apparatus survived the Second World War untouched, and is now stationed at the Institute of Phonetics in Prague. The paper describes in detail the Prague collection of tuning forks, with photographs illustrating the device, and provides a commentary on the use of tuning forks in phonetic research of that time. In addition, the intricate history of the Prague tonometre is discussed, pertaining to the efforts of Josef Chlumský, the founder of the Czech phonetics institute, to obtain the device from the French colleagues.

1. Introduction

Phoneticians have benefited from acoustic developments at least from the year 1863 onwards, when Hermann von Helmholtz published the first edition of his treaty on acoustics and the perception of sound, *Die Lehre von den Tonempfindungen* [1]. Although various devices for acoustic analysis and graphic displays of acoustic parameters have since been constructed (e.g. the oscilloscope, the sound spectrograph, or digital signal processing tools), a phonetician's scientific method still relies to a large part on the human ear. Auditory analysis is as important as acoustic analysis, and the two approaches should ideally complement each other (this is also a premise of many forensic phoneticians; see [2]).

The current paper focuses on the work of the nineteenth-century acoustician, Rudolph Koenig (1832–1903, settled in Paris), who designed and manufactured a great number of acoustic devices that were renowned for their precision and craftsmanship, in addition to performing experiments with these instruments himself. The celebrated Grand Tonometre, of which Koenig made several versions during his life, occupies an exceptional position among the vast array of instruments offered in Koenig's catalogue [3]. The aims of this paper are

- to highlight key events in Koenig's life and work as a precision instrument maker, especially related to the tonometre (mostly based on Pantalony's excellent biography [4]);
- to describe the Prague collection of tuning forks in its present state, including photos of the apparatus, and document the way it was assembled and maintained;
- to provide a commentary on the use of tuning forks in phonetic research.

The study is based on available primary and secondary literature, as well as on archival material of the Institute of Phonetics in Prague. Photographic documentation of the apparatus was carried out by the author.

2. Rudolph Koenig and the Grand Tonometre

2.1. Koenig the instrument maker

The course of Rudolf Koenig's life was to a large extent determined by his upbringing. He was born in 1832 in Königsberg (Prussia), a city renowned for its scientific atmosphere, and his father, a teacher of mathematics and physics, steered young Rudolph towards an interest in science and music. Importantly, among his father's circle of friends there was a certain Hermann von Helmholtz, then a professor in physiology. Nevertheless, failing to complete formal schooling, Koenig decided to move to France and make a living in Paris. He became an apprentice to a celebrated violinmaker, Vuillaume, under whom Koenig stayed for several years and developed the skills that would later become so important – great precision, attention to timbre and the selection of the right material for a harmonically rich sound [4: 6]. What Koenig carried off was above all the passion about his profession, making "work of arts" rather than mere "products for sale" [4: 8].

It was only natural for Koenig to seize the opportunity and aim his attention to the quickly developing field of acoustics. Paris was at the time a centre in the scientific instrument trade (acoustics, optics, electricity), dominating the market. Koenig quickly developed a reputation of a very skilled artisan worker and a reliable business partner, resulting in his unrivalled position as *the* precision acoustical instrument maker from the 1860s until his death forty years later [4]. In this specific trade Koenig simply had no serious competitors; he sold his instruments – always on a personal basis – to a variety of individuals and institutions, including customers in Germany (e.g. Helmholtz), Britain, the USA and Canada. Many important discoveries were thus made possible. As a mark of his internationally prominent status, the French-German manufacturer received a number of awards for his work: a medal of distinction at the 1862 London exhibition, a gold medal at the 1867 Paris exhibition and, above all, a medal of distinction at the famed 1876 Philadelphia Centennial Exhibition [3].

However, Koenig was not only a great instrument maker and inventor, but also a keen experimenter. Part of his work was to conduct experiments in which he measured the precision of his instruments, thereby perfecting his products and surpassing any potential rivals [4: 49]. He knew very well that his business reputation rested on reliability of the instruments and on credit with the scientific community. Although Koenig was a prolific supplier of instruments for Helmholtz, he did not stop there: much of Koenig's time was occupied by disputes with the famous German scientist over the objectivity of combination tones, which Koenig was unable to confirm in his experiments (see [4: Chapter 7] for details). The Germans backed up Helmholtz, but especially the English and the Americans sided with Koenig. This stimulating activity frequently led to innovation or even invention of new instruments on part of Koenig.

Pantalony claims that through his instruments and experiments Koenig transformed the field of acoustics itself [4: 167]. Tuning forks, resonators, wave sirens or manometric flames [3] became a standard apparatus at many universities and physics laboratories [5]. Koenig offered hundreds of instruments, as evidenced by his five catalogues he issued over the years (1859, 1865, 1873, 1882, 1889). He also published extensively throughout his life [6], writing on beats, timbre, vowels, and even ultrasonics, and gave many lectures. Scientists and teachers gathered at his atelier in Paris, sometimes staying at Koenig's apartment for the summer and performing experiments with the expensive instruments [4: 140]. Koenig was the man you went to see first when you swept through Europe procuring equipment for your laboratory. He had many connections, gave advice gladly, and did not strive to enrich himself [7: 13-14].

2.2. The Grand Tonometre

A *tonometre* is simply a set of tuning forks spanning several frequencies. The term was coined at the beginning of the 19th century by Johann Scheibler, a German silk manufacturer with an interest in acoustics, who assembled the first such device [8: 14]. Scheibler's tonometre included 56 tuning forks, covering the range of a single octave from A220 to A440 with a difference of 4 Hz between individual forks. However, it was Koenig who revolutionized the device. As noted above, Koenig produced precision acoustical instruments of excellent quality, which applies to tuning forks as well. He manufactured thousands of them during his lifetime, and he gained invaluable experience by daily work with the various instruments and by running experiments. It can be said without exaggeration that the tuning work, as created in Koenig's acoustical workshop, became "the fundamental instrument of nineteenth-century acoustics" [4: 83], symbolizing the "analytic, elemental theory of sound" [4: 55-56].

Constructing a tuning fork was a very minute and time-consuming process. First, the right (i.e., soft) steel had to be selected so that the sound would be as pure as possible, containing few harmonics. Koenig had to examine its structure, find any cracks in it, and perform experiments regarding its shape and temperature [4: 92-95]. After a rough blank, prepared by Koenig or his assistants, the forks were fine-tuned by hand, which Koenig almost always did by himself as the most critical phase in the process [4: 92]. Even small pieces of steel (less than one millimetre) filed off at various parts of the prongs could make a huge difference. The length and width of the prongs, together with their shape and mass, determined the resulting frequency and quality of the sound [9: 153]. Finally, a tuning fork usually received a wooden resonator box made of pine (see Fig. 1) – Koenig thus amply utilized his violin-making beginnings.

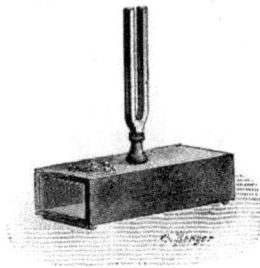

Figure 1: A tuning fork with a resonance box (from the publication of Koenig [3: 21]).

Koenig's exhibition stand was always a great attraction at the international fairs, and much of this popularity was due to the tonometre and some other important instruments for visualizing the qualities of sound (e.g. the phonautograph or the manometric flame capsule; see [6] for the description of the devices). At the London exhibition, Koenig impressed the audience with his first complete tonometre, comprising 65 tuning forks and covering a single octave in 4-Hz steps [4: 55]. The tonometre was a major success and Koenig incorporated it into his catalogue of instruments. At the Paris exhibition, Koenig displayed a more advanced version with 330 tuning forks ranging from 16 Hz to 2,048 Hz [4: 91]. Although he sold many of the smaller instruments, including some tuning forks, the complete tonometre was too expensive even for the Americans [4: 75-77]. The most famous version of the apparatus, the Grand Tonometre, appeared at the 1876 Philadelphia exposition. It consisted of 692 precisely built tuning forks (an amazing number), which corresponded to more than 800 tones as some of the lower frequency forks had adjustable sliders [4: 91]. The total range of frequencies amounted to 16 Hz – 4,096 Hz. Being even more expensive than the previous version, Koenig experienced tremendous difficulties selling the apparatus (and other instruments as well); the tonometre eventually ended up at the US Military Academy at West Point [4: 124-126]. It is presently placed at the Smithsonian Institution in Washington, D.C. (see a photo in [10]).

Koenig immediately began constructing a new tonometre with substantially higher ambitions. It took him decades to complete, until 1894. Although the apparatus included only 158 tuning forks, these were mostly adjustable with sliding weights so that the number of tones represented was more than ten times as many, ranging from 16 Hz to 21,845 Hz [4: 139]. Towards the end of his life, Koenig thus managed to build "the definitive instrument for precision tuning that offered a full range of sounds in the smallest possible gradations of pitch" [4: 140]. Moreover, as he extended the upper range even higher – up to the unbelievable 90,000 Hz [7: 196], [11] – and conducted experiments with the apparatus, Koenig seems to be the first person to measure and record the ultrasound [4: 160]. The *grand tonomètre universel* is shown in Figure 2.

Figure 2: Koenig's *grand tonomètre universel* in the laboratory of Rousselot (from Hála [12: 62]).

3. The Prague collection of tuning forks

3.1. Description of the device

The Prague collection of tuning forks was intended to be a faithful copy of Koenig's universal tonometre. However, the range of frequencies never reached the 90 kHz of the original. Available information ([12: 62] and an inventory of tuning forks in our archives) confirms that in the 1940s there were in total 99 tuning forks covering the frequencies from 140 Hz to 3296 Hz; except for some of the lower forks missing, this is also the current state of the tonometre (Figure 3). Like the Parisian model, the Prague tonometre was mounted on a wooden construction in which the tuning forks were arranged, in several levels, from the lowest (left) to the highest (right) frequencies. In the 1990s, the tonometre also included a certain number of forks above the 3296-Hz limit, but the precise values are not known. These were located in the currently empty uppermost level (Fig. 3). The largest tuning fork in the collection, with the frequency of 16 Hz (adjustable), is 137 cm high and has a special metal stand (Fig. 4).

Since the frequency of each tuning fork was adjustable – with the help of small sliders – to several values, the researcher was able to produce all frequencies within the range of the tonometre in calibrated increments of 4 Hz. By way of example, Figure 5 depicts a 1600-Hz tuning fork the frequency of which can easily be set to the values of 1568, 1572, 1576, 1580, 1584, 1588, 1592 or 1596 Hz if we add the sliders and position them accordingly.

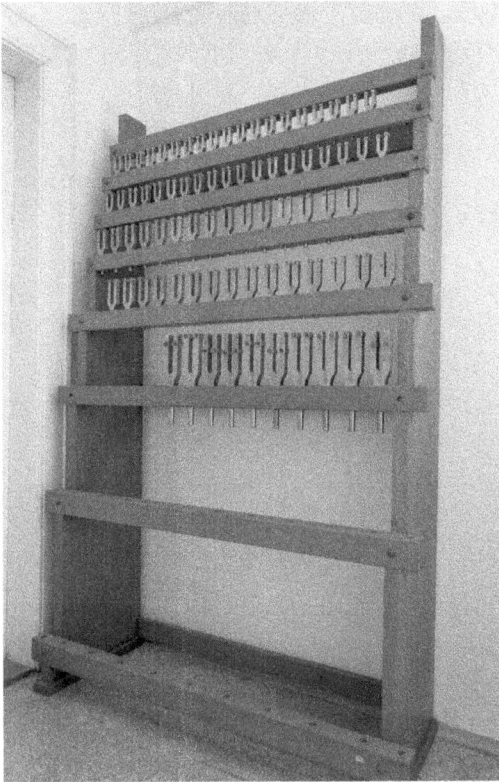

Figure 3: The Prague copy of Koenig's tonometre in 2015 (photo by author).

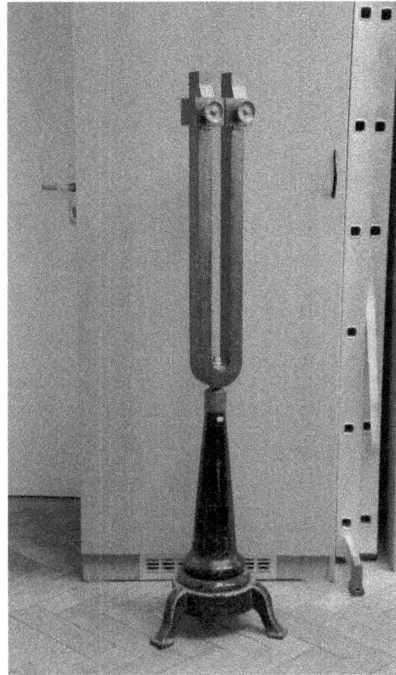

Figure 4: The largest tuning fork (16 Hz, set to 40 Hz) in the collection (photo by author).

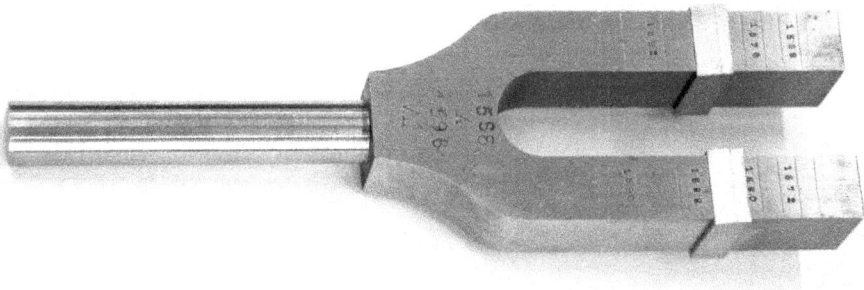

Figure 5: 1600-Hz tuning fork with adjustable sliders from 1568 Hz to 1596 Hz (photo by author).

All tuning forks are made of fine steel and were manufactured in Paris by Henry Lepaute. Thanks to the high craftsmanship, the forks are capable of producing almost pure tones with no significant overtones (see the spectral peaks in Figure 6; one second after striking a 1220-Hz fork there was a relatively strong first harmonics, but it nearly disappeared after two more seconds). The vibrations are strong and last for a long time (up to a minute) when the tuning fork is placed in a wooden resonator box and struck with a rubber hammer (Fig. 7). Moreover, the quality of the material ensures that the characteristics of the tuning forks are stable; after decades of use and then storage, the frequency of the forks did not shift much (only by 5 Hz and 14 Hz for the two forks in Fig. 6).

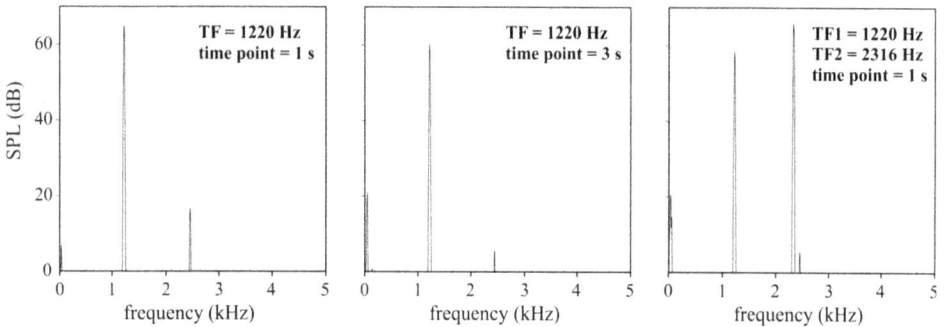

Figure 6: Spectrum of (1) a 1220-Hz tuning fork measured one second after initiation; (2) the same fork measured three seconds after initiation; (3) two simultaneous forks, 1220 Hz and 2316 Hz, measured one second after initiation of the second fork.

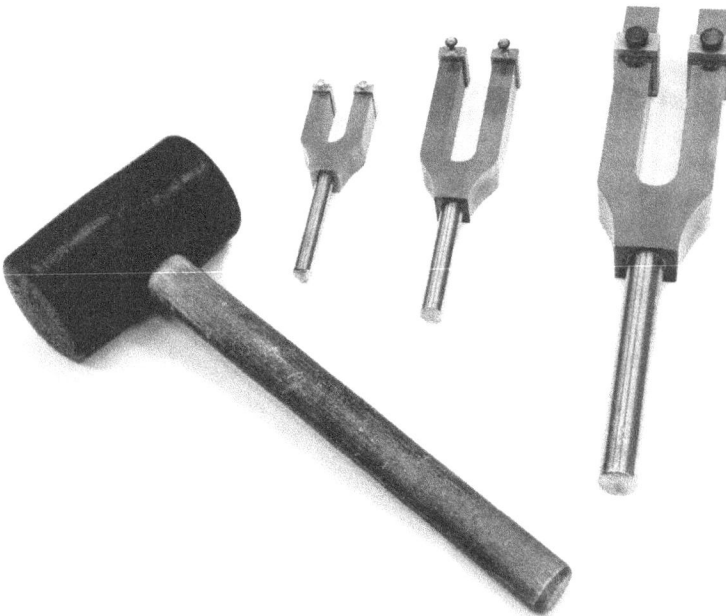

Figure 7: A rubber hammer with three tuning forks: 3296 Hz, 1152 Hz and 774 Hz (from left to right). Photo by author.

3.2. Assembling and maintaining the Prague collection

This section surveys the sources of the Koenig apparatus in Paris and its establishment in the Prague laboratory. Towards the end of his life, Koenig created a universal tonometre which he was planning to sell to the experimental phonetician, Abbé Rousselot, whom he befriended [7: 195]. However, after his death, there were several interested parties involved, including Rousselot, but none of them could afford to pay the full price [4: 142], [7: 195]. Fortunately, Rousselot managed to get a reduced price for the apparatus and procure this still substantial sum from the Collège de France. The long-desired tonometre was transported to his laboratory.

The Czech connection arose through Josef Chlumský (1871–1939), who stayed at Rousselot's laboratory for several years as part of his training. Chlumský proved to have great aptitude for the phonetic science and became Rousselot's most prominent pupil, his "right hand" (Rousselot's personal dedication to Chlumský). He was even reckoned with as Rousselot's successor, but this came to no avail [13: 5]. Chlumský, very well acquainted with the new experimental methods, resolved to build a similar laboratory in Prague after his return from France in 1914. The laboratory was officially established at Charles University in 1919.

According to several reports [13], [14], Chlumský immediately began securing equipment for the laboratory. Due to its high price, the tonometre could be assembled only gradually; it was financed by special subsidies from the Ministry of Education. Entries in the book of transactions preserved in our archive reveal how the tuning forks were acquired between the years 1921 and 1931. The accounting demonstrates the money that was spent on:

- *tuning forks*: e.g., a set of 13 large tuning forks was bought in 1922 for a total of 4,225 francs, which equalled 24,230 Kč (Czechoslovakian crowns) at the time; in today's currency, it roughly corresponds to the sum of 700,000 Kč (Czech crowns), or 25,500 euros;
- *commissioned work and various accessories*: with regard to the same set of tuning forks, the calibration, filing, frequency information engraving and adding sliding weights cost nearly as much as the forks themselves (19,728 Czechoslovakian crowns ≈ 570,500 Czech crowns ≈ 20,700 euros);
- *delivery costs*: the boxes were heavy; e.g., the 13 forks were transported in five crates (with other equipment) for the sum of 1,970 Kč (≈ 57,000 Kč today ≈ 2,100 euros);
- *travel expenses*: Chlumský personally went to France to commission orders and also to fine-tune several tuning forks to reduce the price [13], [14].

The entire collection of tuning forks is valued at about 122,000 Czechoslovakian crowns (≈ 3,500,000 Czech crowns ≈ 128,280 euros). However, it must be noted that the conversion is a very rough and inexpert estimate based on a comparison of the monthly income of an office worker in 1930 and in 2015.

In 1931 the laboratory moved to a new faculty building, where it remained ever since, so it was finally possible to find adequate accommodation for the tonometre. There were 95 tuning forks in 1932 [14: 2] and 99 in 1941 [12: 62]. The tonometre is thought to have comprised one additional row of small tuning forks but no written records presumably exist to establish the number or frequencies of these items. However, the wooden construction for the tonometre does indeed contain the additional level, with appropriate holes in it (see Fig. 3). Assuming that there are on average three tuning forks in each one hundred Hertz, the tonometre could theoretically have reached up to approximately 4,000 Hz.

Since the reopening of universities after the Second World War, the apparatus – which magically survived untouched – was kept in the director's office (i.e., that of Bohuslav Hála), protected by a shield of glass against dust. Hála had used the tuning forks for his study of vowel acoustics [12], but since the 1950s the collection had a status of a historical device. As one of the contemporaries recollects (Z. Palková, 2015, personal communication), interested students could examine the collection if they wanted to see it, for instance after a lecture which mentioned tuning forks, but no one was allowed to touch it without leather gloves. The collection was relocated to a different room after the director's retirement in the 1960s, and was later placed in the corridor for a public display where it stayed until 2011. Sadly, about 20 tuning forks (the heaviest pieces representing the lowest frequencies) were stolen in 2004 or 2005 by an unknown perpetrator. Although the remaining tuning forks are kept in a storeroom at the moment, the Institute of Phonetics has immediate plans to arrange a permanent exhibition in the corridor again. We also hope to create some replacement for the missing forks to fill the embarrassingly empty places in the two lowest rows.

3.3. Tuning forks in phonetic research

The tuning fork was not only an indispensable instrument in acoustics, but also in early phonetics. Tuning forks were used in a number of ways in phonetic research.

Perhaps the simplest case was at the same time most important. Tuning forks were used to *measure the duration of speech sounds*. Chlumský [15] reports that this was common practice at the laboratories in Paris and, by extension, in Prague. A 200-Hz tuning fork was connected to the kymograph and its vibrations were recorded along with the speech sample. Two periods produced by the tuning fork therefore correspond to 10 milliseconds (Fig. 8). This procedure was employed in a wide variety of studies of that time, and kymograph tracings are accompanied in many reports by vibrations of a tuning fork ([6], [16], [17], [18], [19], [20], [21]).

Tuning forks can be used as a tool that helps the experimenter *estimate the pitch of a tone*. This is a straightforward process. One could take the piano for the same purpose, but tuning forks are much better suited since they offer a finer scale of frequencies (differences of only a few Hertz) and produce few harmonics. The experimenter simply compares the pitch of the given sound to be analyzed with various tuning forks and selects the best fit.

However, other uses of the instrument in phonetic research require substantial training. Rousselot remarks on this in his seminal book [5: 165], and Hála likewise stresses that "working with them is difficult" [12: 269]. In the hands of an experienced researcher, tuning forks can be used as a tool for the *investigation of vowel resonances (formants)*. In vowels, the vocal folds produce a complex tone characterized by the fundamental frequency (F0) and its upper harmonics. This source signal is then shaped by the cavities and articulatory organs above the larynx. Since different vowels have different settings of these articulatory parameters (e.g. tongue position), the resonance characteristics of the supralaryngeal system change accordingly. The fundamental fact is that F0 and formant frequencies are independent of each other. Therefore, it is also possible to investigate vowel resonances in whispered speech despite the lower amplitude of the signal.

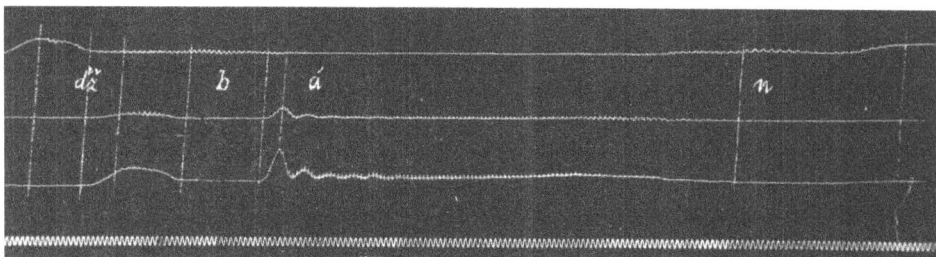

Figure 8: Kymograph engraving showing a tuning fork of 200 Hz at the bottom as a measure of time (from the original publication of Chlumský [15: 19]).

Specifically, the investigator places a tuning fork (or a whole set of tuning forks) in front of the speaker's mouth and examines which of the forks are forced to vibrate as a result of the sound (Figure 9; [11], [12], [22]). If the particular frequency is present in the vowel, the tuning fork will start to vibrate. Importantly, the experimenter pays attention to the force with which the tuning fork responds, comparing it across different frequency bands. If the response is, at a specific region, markedly greater than at other regions, it corresponds to a formant of the given vowel. However, it is possible to investigate only the first three resonances (plus F0), the effect is otherwise quite weak [12: 61]. In addition, tuning forks can be supplemented with resonators to reinforce specific frequency bands.

Figure 9: Determining formant resonances by means of a tuning fork
(from the original publication of Rousselot [22: 23]).

Tuning forks were employed, for instance, by Helmholtz [1] and Koenig [6] to determine vowel resonances for German, or by Rousselot [11], [23] for French. Bohuslav Hála conducted an extensive study of Czech vowels, which was published in 1941 as *Acoustical Basis of Vowels* [12], a book of more than 300 pages. Hála examined Czech vowels by means of auditory analysis, tuning forks and resonators and mathematically by computing harmonic Fourier analyses of the waveform. All these methods were exacting and extremely time-consuming, so it took several years before the research was completed. Hála had to determine the frequencies and amplitudes of formant resonances in different types of material (words, sentences, connected speech) produced by four speakers. It is notable that despite the limited technical possibilities Hála was able to describe the formant values quite accurately (cf. [24]).

The forth use of tuning forks pertains to *vowel synthesis*. By choosing the right set of forks with appropriate amplitudes of the individual frequencies, one can synthesize, from pure tones, the desired vowel qualities [5: 166-168]. Rousselot [5: 167] includes a drawing of Helmholtz's electrically-operated synthesizer improved and created by Koenig, which uses tuning forks to provide harmonics starting from 128 Hz, and a set of resonators that are manually adjustable for sensitivity to the excitation from the vibrating tuning forks, thus simulating the resonances. See also the description in Helmholtz [1: 120-123] or in [4: 32].

Finally, tuning forks were used to *estimate the auditory range of the listener* [11: 23-24], [23: 11]. Since Koenig managed to manufacture forks exceeding the upper limit of hearing, both patients and healthy subjects can be screened for their ability to detect tones presented to them. Similarly, Hála [12: 120] employed tuning forks for investigating the *perceptual sensitivity to individual frequencies* by measuring the distance to which the sound of a given tuning fork, brought into motion by a constant force across trials, was still perceptible to the listeners. The results confirmed that the ear was not uniform in its sensitivity within the range. He conducted the experiment in a large forest clearing under favourable weather conditions in 1933.

4. Conclusion

The purpose of this article was to describe the Prague collection of tuning forks, a legacy more than a hundred years old. Koenig's final version of his universal tonometre, covering the frequencies between 16 Hz and 90,000 Hz, was purchased by the Parisian experimentalist Abbé Rousselot in 1901. Josef Chlumský, Rousselot's student and collaborator for several years, returned to Prague with a vision to establish a laboratory of experimental phonetics similar to the one he had grown to love in Paris. Over the years, starting from 1921, Chlumský commissioned and gradually procured a faithful copy of Koenig's tonometre (only with a smaller number of tuning forks). The Prague tonometre has survived until today, although a part of it was, sadly, misappropriated. In addition to the written and photographical description of the device, it has also been shown here how tuning forks can – or used to be – employed in phonetic research. Our main argument is that despite the far-reaching development in technology, phonetic experimentalists at the beginning of the twentieth century made significant contributions to the field even with relatively simple and less accurate instruments. Bohuslav Hála's book on the acoustics of Czech vowels, drawing mainly on auditory analysis and on work with tuning forks and resonators, was ground-breaking in its scope and thoroughness, and, by and large, remained unparalleled ever since (needless to say, new descriptions of formant values, based on a larger population and contemporary material, will be more useful to modern researchers; this, however, does not lessen the significance of Hála's work). The historical background regarding R. Koenig and the early researchers helped to frame the Czech situation within a larger, European perspective.

Acknowledgements

This research was supported by the Charles University in Prague programme for science development P10-Linguistics, and by the Faculty of Arts development project *History of Phonetics – course innovation* awarded to the author. I would also like to thank my colleagues at the Institute of Phonetics for sharing with me their thoughts and memories of days gone by.

References

[1]   von Helmholtz, H.: *On the Sensations of Tone [English 1885 translation by A. J. Ellis]*. New York: Dover, 1998.

[2]   Künzel, H. J.: Current approaches to forensic speaker recognition. *Proceedings of ESCA Workshop on Automatic Speaker Recognition, Identification, and Verification* (1994), 135–141.

[3]   Koenig, R.: *Catalogue des Appareils d'Acoustique construits par Rudolph Koenig*. Paris, 1889.

[4]   Pantalony, D.: *Altered Sensations: Rudolph Koenig's Acoustical Workshop in Nineteenth-Century Paris*. Dordrecht: Springer, 2009.

[5]   Rousselot, P.-J.: *Principes de Phonétique Expérimentale [Vol. I & II, revised edition]*. Paris: Didier, 1924.

[6]   Koenig, R.: *Quelques expériences d'acoustique*. Paris: Imprimerie A. Lahure, 1882.

[7]   Rousselot, P.-J.: *Phonétique expérimentale et surdité*. Paris: Institut de laryngologie et orthophonie, 1903.

[8]   Rossing, T. (ed.): *Springer Handbook of Acoustics [2nd edition]*. Dordrecht: Springer, 2014.

[9]   Mehnert, D.: *Historische phonetische Geräte: Katalog der historischen akustisch-phonetischen Sammlung (HAPS) der Technischen Universität Dresden*. Dresden: TUDpress, 2012.

[10]   Smithsonian Institution, National Museum of American History: Tuning Forks. Accessed 8.7.2015 at <http://americanhistory.si.edu/science/tuningfork.htm>.

[11]   Rousselot, P.-J.: Sur les caractéristiques des voyelles, les gammes vocaliques et leurs intervalles. *Comptes-rendus de l'Academie des Sciences* 137 (1903), 40–43.

[12]   Hála, B.: *Akustická podstata samohlásek [Acoustical Basis of Vowels]*. Praha: Česká akademie věd a umění, 1941.

[13]   Hála, B.: *Josef Chlumský*. Praha: Česká akademie věd a umění, 1940.

[14]   Janko, J.: Několik slov o životě a působení Josefa Chlumského [A few words about the life and activities of Josef Chlumský]. *Časopis pro moderní filologii* 17 (1931), 1–5.

[15]   Chlumský, J.: *Pokus o měření českých zvuků a slabik v řeči souvislé [An attempt at measuring Czech sounds and syllables in connected speech]*. Praha: Česká akademie císaře Františka Josefa pro vědy, slovesnost a umění, 1911.

[16]   Rousselot, P.-J.: Phonétique d'un groupe d'aïnos. In: Rousselot, P.-J., Pernot, H. (eds.), *Revue de Phonétique [vol. 2]*. Paris, 1912, 5–49.

[17]   Rousselot, P.-J.: Dictionnaire de la prononciation française. In: Rousselot, P.-J., Pernot, H. (eds.), *Revue de Phonétique [vol. 2]*. Paris, 1912, 159–191.

[18]   Chlumský, J.: La question du passage des sons. In: Rousselot, J.-P. and Pernot, H. (eds.), *Revue de Phonétique [vol. 2]*. Paris, 1912, 80–93.

[19]   Chlumský, J.: *Česká kvantita, melodie a přízvuk [Czech quantity, melody and accent]*. Praha: Česká akademie věd a umění, 1928.

[20]   Šrámek, E.: Les consonnes rétroflexes du Bengali. In: Pernot, H. (ed.), *Revue de Phonétique [vol. 5]*. Paris: Didier, 1928, 206–259.

[21]   Millet, A.: Phonétique et chirurgie. *Časopis pro moderní filologii* 17 (1931), 106–111.

[22]   Rousselot, P.-J.: Classification des voyelles orales. In: Rousselot, P.-J.. Pernot, H. (eds.), *Revue de Phonétique [vol. 1]*. Paris, 1911, 17–32.

[23]   Rousselot, P.-J.: *Phonétique expérimentale et surdité*. Paris: Institut de laryngologie et orthophonie, 1903.

[24]   Skarnitzl, R.; Volín, J.: Referenční hodnoty vokalických formantů pro mladé dospělé mluvčí standardní češtiny [Referential values of vocalic formants for young adult speakers of Standard Czech]. *Akustické listy* 18 (2012), 7–11.

William Holder – A Pioneer of Phonetics

Angelika Braun

University of Trier, Germany
brauna@uni-trier.de

Abstract: This paper looks at William Holder's "Elements of Speech" [1] from the perspective of modern phonetic and phonological literature. The structure of this presentation, first touching on general linguistic concepts, follows the three principal stages of speech production: respiration/air stream mechanism, phonation, and articulation. Holder's rendition of speech perception is discussed in conjunction with the clinical application of teaching a deaf-mute to speak. Furthermore, the passage on what is today called prosody will be reviewed. The mere fact that this topic is addressed in Holder's work is remarkable since suprasegmentals were not normally an issue at all in 17th century publications on speech. The extent and precision of the description in the "Elements of Speech" is truly remarkable. A comparison with the writings of Wolfgang von Kempelen, who is often referred to as *the* pioneer of modern day phonetic theory, shows that Holder knew much more about speech and hearing than Kempelen even though his writings have received much less attention.

1   Introduction

The history of what is today called phonetics has received relatively little attention. As Abercrombie [2: 1] puts it, "There is, in fact, a firm opinion in some quarters that before about 1830 there was no such thing as phonetics." However, while the term "phonetics" may not have been around at that time, the working principles of the speech and hearing mechanisms were certainly described in various contexts: anatomical, orthoepic, linguistic, and clinical, to name only a few. Thus, the writings of those early days of speech analysis can be assessed in view of present-day knowledge.

This contribution focuses on a 17th century publication by William Holder entitled *Elements of Speech. An Essay of Inquiry into the Natural Production of Letters: with An Appendix Concerning Persons Deaf & Dumb*. So far, the focus of the scientific discussion on Holder's writings has been on his musicological works and, as far as speech is concerned, writings which are related to teaching a deaf-mute speaker to talk [3, 4]. Some authors emphasize his contribution to general phonetic knowledge [5, 6, 7], but somehow the full scope of the modernism of his approach does not seem to have been recognized.

Instead, Wolfgang von Kempelen has been credited with pioneering work on the analysis and synthesis of the speech process : "Den Grundstein dazu [i.e. die für das Zustandekommen eines Lautes nötige subtile Arbeit in allen ihren Einzelheiten zu ergründen, A.B.]" legte 1791 v. Kempelen, der durch seine erstaunliche Analyse und Synthese des Sprechmechanismus diese neue Richtung der phonetischen Forschung eröffnete, eine Leistung, die, fachgeschichtlich betrachtet, noch immer nicht gebührend eingeschätzt wurde [...]"[1] [8: 161].

---

[1] The foundation for this (i.e. for establishing the subtle detail which is required to constitute what is called sound) was set by Wolfgang von Kempelen [9], who, by way of his amazing analysis and synthesis of the

2   The Man

William Holder was born in 1616 and studied at Pembroke Hall, Cambridge, where he received an M.A. and became a fellow in 1640. He received his Doctor of Divinity from Oxford in 1662 and was elected Fellow of the Royal Society in 1663. He is primarily remembered as a musicologist, having published "A Treatise on the Natural Grounds and Principles of Harmony" in 1694.

A controversy arose within the Royal Society about his claim to have taught a deaf-mute called Alexander Popham to speak in 1659. That is why his book, which the present contribution is based on, contains an appendix "Concerning Persons Deaf and Dumb", in which he describes the method employed. In order to better understand the relevance of this issue it has to be pointed out that teaching a congenitally deaf person to speak seems to have been a major issue in the 17th century which was addressed in several monographs (cf. e.g. Amman [10], Bulwer [11]). The case of Alexander Popham evidently became the subject of a bitter dispute within the Royal Society between Holder and John Wallis[2], possibly because Popham suffered a relapse and was subsequently treated by John Wallis, who then claimed responsibility for the success [3: ii]. Regardless, Holder as well as Wallis based their teaching concepts on their respective understanding of the speech production mechanism. Firth [13: 115] insinuates that the publication of Holder's work was sabotaged by his rivals, possibly because it was more "modern" than they deemed appropriate.

3   The Framework

In order to gain a better understanding of the innovative force which is contained in Holder's writings, we should first consider the framework in which Holder and his generation were operating. Specifically, the sources of information on the physiology of speech and hearing need to be taken into account.

In antiquity, the Greek physician Galenus was one of the very few who gained detailed knowledge about voice and speech production by dissecting human cadavers [8: 201]. During the Middle Ages, a ban on autopsies was quite rigorously enforced by the Catholic Church. This prevented researchers from exploring the anatomical bases of speech production and resulted in rather gross ideas about them [8: 73; 202]. During the Renaissance, scholars like Michelangelo Buonarotti and Leonardo da Vinci (secretly) performed autopsies in Italy and thus gained insight into the anatomy and physiology of the voice and speech production mechanisms (cf. the da Vinci anatomical sketches and [15: 42].

Thus, 16th and 17th century scholars as e.g. Amman [16] and Holder had relatively little to go on for their descriptions of speech production. It is therefore commendable that Holder would make the distinction between respiration, phonation and articulation, which is common today but was by no means well-established in the 17[th] century. If we compare Holder's writings to other 17[th] or even 18[th] century literature (e.g. Christopher Cooper, Alexander Hume, Petrus Montanis, or Francis Lod[o]wick, cf. [5: 2; 12: 357]), we cannot but admire his insights and

---

speech mechanism, founded this new branch of phonetic sciences; an achievement which has still not been fully appreciated to this date.
[2] John Wallis (1616-1703) who taught the same patient and authored a monograph entitled "Grammatica Linguae Anglicanae" [with an appendix: De loquela, sive sonorum formatione, tractus grammatico-physicus.] London: Leon Lichfield [13].

the "modern" approach to the subject matter (see also [5]). This is an asset which is emphasized by Firth [14] who states that "[h]is descriptions of the sounds are quite good even judged by present-day standards" (p. 115). The following summary of Holder's work is organized along the lines of a present-day textbook of phonetics.

4 Holder on Letters and Sounds (and the Concept of Phonemes)

At first glance, it seems that – very much in keeping with other writings of the time – Holder confounded the concepts of letters and speech sounds. This then would have constituted one of the few notions in Holder's writings which are no longer generally acceptable today. Abercrombie [5: 317], however, points out the ambiguity of the term "letter" and traces it back to the 17th century: "'An Element of Speech, which is commonly call'd a Letter, hath a double signification' – it could be something you hear or something you see'" [17: 1].

Holder does in fact draw a clear distinction between "letters" on the one hand and "symbols" or "characters" on the other (p. 12): "[...] the incongruous pronunciations of several Letters, as they lie described to the Eye by Symbols or Characters of the Alphabet of several Languages, which indeed ought to be only one [...]"(ibid.). This translates his "letters" into constituting what are today called "sounds" (or even phonemes) and his "symbols" or "characters" constituting what are today called "letters". He later talks about "Signs *Audible*" and "Signs *Visible*" and calls for alphabets in which "[c]haracters or Signs written were exactly accommodated to Speech" (p. 108), which makes it clear that Holder was well aware of the difference between sounds and letters.

Further evidence for a phonemic perspective can be found later in his book, when Holder writes: "There is so much space between *a* and *e*, that there may be a vowel inserted between them, and a fit character for it may be *æ*, and perhaps **some Languages may have a distinct use of such a vowel**" (p. 81; emphasis mine, A.B.). The concept of distinct use of vowels, which is language-specific, clearly points to the phonemic principle. Incidentally, the Ash-symbol (A-E ligature) [æ] has made its way into the International Phonetic Alphabet, denoting precisely the vowel quality which Holder described.

5 Holder on Speech Production

Today, it seems so perfectly natural to talk about speech production in terms of respiration/air stream mechanisms, phonation, and articulation ([18, 19, and 20], to cite only a few) that we have become oblivious of the fact that this descriptive framework seems to have been lost in the 18th and early 19th centuries. It would seem that only in the 19th century did the knowledge about speech production fully return [15: 202; 21: 6].

5.1 The Source-Filter Theory of Speech Production

This theoretical concept is one of the true basics in speech production which can be found in any current textbook. This model is generally attributed to Gunnar Fant [22] who described the speech signal as the complex output of laryngeal activity (source) and supralaryngeal processes (filter).

Interestingly enough, Holder draws a similar distinction: He talks about "material" and "formal" causes of what he calls letters (i.e. sounds, see above): "Their *Matter* is various; *viz.*

Breath, or voice, *i.e.* Breath vocalized by the operation of the *Larynx*. Their *Form* is constituted by the Motions and Figures of the Organs of speech, affecting the Breath or Voice with a peculiar sound, by which each Letter is discriminated" (p. 64).

This distinction between matter and form, where "matter" refers to laryngeal activity and "form"[3] to supralaryngeal activity, clearly reflects the distinction between source and filter as outlined in Fant [22]. The description provided by Holder is, of course, not based on acoustical data, but its analytical precision is still remarkable.

5.2   Respiration

Holder limits his remarks on the role of the lungs to a short description: "The *Lungs* are as *Bellows*, which supply a force of Breath: the *Aspera Arteria* is as the nose of *Bellows*, or as a channel in the sound Board of an Organ, to collect and conveigh the Breath, and somewhat more, by a power of contracting and dilating it self, which those have not" (pp. 22-23). This can be taken to imply that Holder is exclusively considering an egressive pulmonic airstream mechanism. This is in keeping with the focus on the sounds of English, which phonologically makes exclusive use of egressive pulmonic speech sounds. The description of the lungs as bellows has prevailed to this day (cf. Raphael et al. [23: 56]).

5.3   Phonation

Holder is obviously well aware of the fact that the process of voicing happens in the larynx. However, he fails to recognize the nature of the various tissues involved in voice production. Instead, he considers "a vibration of [...] Cartilaginous Bodies" to "form[s] that Breath, into a Vocal sound or Voice [...]" (p. 23). Holder thus was oblivious to the essential role of the vocal folds in voice production. Subbiondo [7], in expressly commending the fact that Holder "knew that the vocal chords were 'cartilaginous bodies'" (p. 174), fails to notice this inaccuracy in his rendition of the phonation process.[4] – His description of the vocal folds as being cartilaginous, which is, of course, partly, but not entirely accurate, does not keep Holder from focussing on the sole difference between, e.g., /p, t, k/ and /b, d, g/ as one between "Articulations of Breath [...] and Articulations of Voice, or Breath vocalized" (p. 38). He goes into quite a bit of detail, characterizing /p/ as "wholly *Mute*, because it is nothing but Breath stopt" (pp. 38-39) and /b/ as accompanied by "a *murmuring* sound of the Voice, formed in the *Larynx*, and passing till it be stopt by the Appulse of the Lips" (p. 39).

5.4   Articulation

Holder's definition of "Articulation" is a very modern one: "By *Articulation* I mean a peculiar Motion and Figure of some parts belonging to the Mouth between the Throat and Lips, whereof some are more easie to be discerned and described [...]. Most [difficult] of all [are] the Vowels, where there are peculiar Figures of the Cavity of the Mouth between the Tongue and the Arch of the Palate[...]" (p. 27). He then proceeds to classify vowels as opening sounds and consonants as sounds involving some kind of obstruction:

---

[3] For a detailed discussion of these concepts cf. Abercrombie [21].
[4] Abercrombie [21] equates the notion of "cartilaginous Bodies" with "the sides of the larynx" (p. 4) and thus seems to accept it as an adequate description of voice production.

"[...] That in all *Vowels* the passage of the mouth is open and free, without any appulse[5] of an Organ of Speech to another: But in all *Consonants*, there is an Appulse of the Organs, some-times (if you abstract the Consonants from the Vowels) wholly precluding all sound; and in all of them, more or less, checking and abating it" (p. 29). This could well be a description contained in a 21st century textbook.

5.4.1   Consonants

Holder classifies the consonants by
-   articulation of voice vs. articulation of breath (i.e., voiced vs. voiceless [p. 23])
-   appulse (plenary/occluse/close vs. partial/pervious [p. 36])
-   place of articulation (pp. 37-39).

This, of course, reminds the reader of Abercrombie's three-term labels [18: 52] and the classification of consonants in basically every textbook of our time. – In this section, Holder discusses the distinction between nasal and oral consonants in a highly specified way which is totally acceptable from a present day point of view. First, he mentions the various places of articulation and the sounds thereby made. He then proceeds to specify the difference between (oral) stops and nasals:

"Thus the same Articulation; if of *Breath*, makes one letter; if of *Breath vocalized*, or voice, another; If of *voice Nasall* (i.e. when the *Uvula* is opened, and the voice passeth into the Mouth, and is there Articulated, and at the same time hath a free passage through the Nose) then it makes another; and lastly, if of *Breath Nasal*, then another" (pp. 33-34).

These categories may well be interpreted within the framework of distinctive feature theory, *vocal* vs. *spiral* corresponding to +/- voiced, and *naso-vocal* vs. *naso-spiral* corresponding to +/- nasal.

The remaining descriptions are non-binary and look very similar to those provided by Lade-foged [19] or Catford [20]. The following places of articulation are mentioned: labial, labiadental, linguadental, gingival, and palatick. These roughly correspond to the modern-day labels bilabial, labiodental, dental, alveolar, and velar. The degrees of closure are labelled *close appulse* (stops), *pervious appulse* (fricatives), *free* (approximant), and *jarring* (tap/trill). Interestingly enough, Holder's consonant chart (Figure 2, see below) lists the full range of speech sounds, filling all of the cells, amounting to a total of 36. He then proceeds to exclude those speech sounds "that prove not graceful, nor easie to be pronounced, *viz.* 2 *Spiritals*, 9 *Naso-Spiritals*, and 6 *Naso-Vocals*, in all 17; there will remain 19 *Consonants*, proper for use according to the design of Letters" (p. 66; cf. also p. 59). Thus, Holder assumes 19 consonant phonemes.

From today's perspective it must be noted that /h/, /j/, /w/, and /ʔ/ are not listed in the chart. A closer look reveals, however, that Holder addresses all except /j/ in the text. The latter is, however, listed alongside [ʃ] in his figure on page 96 (cf. Fig. 2). As concerns the remaining phonemes, Holder defines the glottal stop as "one stop, whichmay [sic] be made in the *Lar-ynx*, of Breath, before it comes to the Tongue and Palat" (p. 60). He also gives a reason for not listing it as a phoneme 'letter': "[...] I thought it not worthy to be inserted amongst the Letters

---

[5] Meaning "closure" or "narrowing". Holder later introduces "close appulse" to designate plosives and "pervious appulse" to designate fricatives (p. 62).

in that it is applyed to Breath immediately as it passeth through the *Aspera Arteria*, and not to Breath or Voice *Oral* or *Nasal* (pp. 72-73).

As far as /h/ is concerned, Holder remarks that "[...] H is onely a *Guttural Aspiration, i.e.* a more forcible impulse of Breath from the Lungs, applyed when we please, before or after other Letters" (p. 67). He concludes that it "[...] cannot properly be called a Letter, according to that description we have made of Letters; yet in that it causes a sensible, and not incommodious Discrimination of Sound, it ought to be annexed to the *Alphabet* (pp. 68-69).

The phoneme /w/ is mentioned only in passing, i.e. in conjunction with the pronunciation of /h/: "In WHAT, WHICH, and the like, H is pronounced before W. and so of right ought to be written" (p. 72). This evidently refers to the voiceless labial-velar fricative [ʍ].

Articulations	Spirital	Vocal	Nafo-Spirital	Nafo-vocal	
Labial	P	B	+M'	M	3
Oingival	T	D	+N	N	3
Palatick	K	G	+Nɡ'	Nɡ	3
Labiadental	F	V	+F	+V	2
Lingua-dental	Th	Dh	+Th	+Dh	2
Gingival-Sibilant	S	Z	+S	+Z	2
Palatick-Sibilant	Sh	Zh	+Sh'	+Zh	2
Gingival-Free	+L'	L	+L'	+L	1
Gingival-jarring	+R'	R	+R'	+R	1
	7	9	0	3	19

Figure 1: Consonant chart (Holder, p. 62)

5.4.2 Vowels

Holder's description of the vowel system is more difficult to interpret because it refers to the Early Modern English state of affairs. The pronunciation of the vowel phonemes (and possibly also the phoneme system) were quite different then from what they are now.
In an attempt to put Holder's description of vowels into perspective, various sources on the pronunciation of his time were consulted [24, 25, 26].

Holder, when naming the vowels of his time, pursues a similar path as did John Wells [27] with his lexical sets more than 300 years later by citing what he considers to be typical representations of the respective vowels. The following table represents the lexical items as listed by Holder, his notation, and the inferred IPA symbols.

Table 1: Lexical items as listed and transcribed by Holder

Lexical Item	Holder's notation	IPA notation
Fall	ɑ	ɑː/ɔː
Fate	ä	æː/ɛː
Seal	ë	eː
Eel	i	iː
Cole	ò	oː
Fool	òo	uː
Rule	u	iuː
Two	ɣ	oʻː
Folly	[6]	o/ɔ
Fat	[6]	æ
Sell	[6]	e/ɛ
Ill	[6]	ɪ
Full	[6]	ʊ

As previously mentioned (cf. chapter 4 above), Holder proposes to insert a vowel *æ* between *a* and *e*. The examples which are given may not be exhaustive from a present-day perspective, but they represent the state of affairs in the second half of the 17th century.

Holder defines vowels as sounds "made by a free passage of *Breath Vocalized* [i.e. voicing, A.B.] through the cavity of the Mouth, without any appulse of the organs; the said cavity's being differently shaped by the postures of the Throat, Tongue and Lips, some or more of them, but chiefly the Tongue" (p. 80). He also recognizes that, owing to the continuous nature of vowel articulation, the number of phonetic vowels is unlimited, but that the number of vowel phonemes into which these articulations can be grouped is much smaller: "As to the *Number* of Vowels, they, being differenced by the shape of the cavity of the mouth, may be reckon'd very many, if small differences be allowed. But **those which are remarkably distinguished** [emphasis mine, A.B.] [...] may be reduced to these Eight [...]" (pp. 80-81). This clearly points to a phonological approach.

According to Holder, vowels are more difficult to describe than consonants for the following reasons: "The *Articulations*, that is, the Motions and Postures of the Organs in framing the Vowels, are more difficultly discerned, than those of the Consonants; because in the Consonants, the Appulse is more manifest to the sense of Touching, but in the Vowels it is [...] hard to discern the Figures made by the Motions of the Tongue, (inclining onely toward the Palat, and not touching it)" (p. 82). This description of the problem would easily pass for a 21st century textbook rendition. Still, his concept of vowel production is by far less clear and by far less accurate than that of consonant features.

Some notions contained in Holder's book correspond to features which are used today: For instance, his awareness of the dimensions "tongue height", which he calls "aperture" and "tongue position", which he calls "situation" can be inferred from his characterization of /i/ as "being the closest and forwardest" vowel (p. 89). But that scheme is not pursued in a systematic way, the only indication given by Holder is ordering "the Series of the Vowels according to their degrees of aperture, and recess towards the *Larynx*" (p. 90). Thus, he does not draw a

---

[6] No symbol given.

clear separation between the two. He distinguishes four classes of vowels, two of which represent places of articulation and two describe the role of labiality (cf. Figure 2).

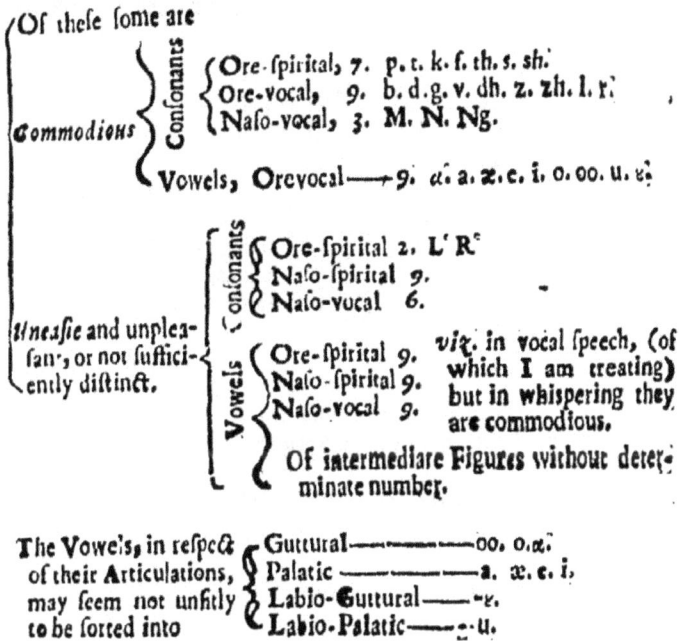

Figure 2: Chart of consonants and vowels with respect to voicing and nasality (Holder, p.98)

The places of articulation for vowels are "guttural" and "palatic", which would translate into front and back in modern terms, and the assignments of the different vowels to these places correspond to the present-day understanding of the distribution. As far as labiality is concerned, the description is less clear: Whereas in his (o), which is also [o:] in terms of IPA, he mentions the role of the lips as "drawing them a little rounder, [which] helps to accomplish the pronunciation of it" but he still concludes that this "is not enough to denominate it a *Labial* Vowel, because it receives not its Articulation from the lips" (p. 86). On the other hand, he lists ɤ and u as "peculiar, [in] that they are framed by a double motion of Organs, that of the Lip, added to that of the Tong; and yet either or them is a single Letter, and not two, because the motions are at the same time, and not successive [...]" (p. 88). This clearly denotes these vowels as labial. Some of his further remarks, however, are much less clear: "Thus u will be onely i Labial, and ɤ will be oo Labial, that is, by adding that motion of the under-Lip, i will become u, and oo will become ɤ"(p. 90). It seems unlikely that this description is explicable by the vowel pronunciation of his time; instead, a more likely cause for this rendition is a misconception of the interplay between tongue position and the role of vowel rounding in general.

On the other hand, it is remarkable that Holder recognizes the physiological and linguistic possibility of nasal and voiceless vowels, but he does not seem to make the connection to their linguistic use in any particular language such as French or Polish. Instead, he classifies them as "*uneasie* and unpleasant" (p. 98).

113

Matter of Sound, ⎰ Oral ⎧ Ore-spirital.
prepared by the ⎱ Breath ⎰ Ore-Nasal. ⎧and may⎫ Nafo-spirital.
Lungs, Larynx, ⎱ Voice. ⎰ Oral ⎩ be filled⎭ Ore-vocal.
Mouth, Nose. ⎱ Ore-Nasal. Nafc-vocal.

Forme, Articulation by

Appulfe of one Organ to another, *Confonants* by degree — Plenary Close:
⎰ Lip to Lip. Labial, as   B.
⎨ Teng to Gums. Gingival,   D.
⎩ Tong to Palat, Palatic,   G.

Partial Pervious:
⎰ Lip to Teeth. Labiodental V.
⎨ Tong to Teeth, Lingua-   Dh.
⎩   (dental

Tong to ⎰ Sibilant.   Z.
Gums,   ⎨ Jarring.   R.
Gingival ⎩ Lateral.   L.

Tong to Palat, Palatic, Zh. or J.

Inclination of one Organ to another without Appulfe.

Vowels — a. a. æ. e. i. o. oo. u. u.

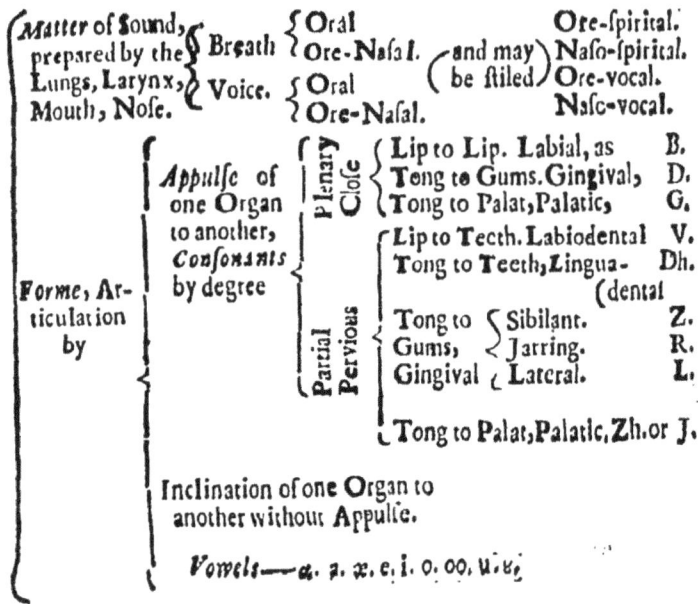

Figure 3: Comprehensive chart of consonants and vowels (Holder, p.96)

It is evident from Figure 3 that Holder ventures to incorporate vowels and consonants into one single chart. The vowels in this chart are distinguished from other sounds by what would today be called the feature +/- approximant [28]. This means that to some extent at least he applies descriptive parameters to consonants and vowels alike, once again a very modern concept of looking at speech sounds.

In addition to the vowels, Holder also mentions the notion of diphthongs. However, he does not regard the combination of e and o with a as "proper *Diphthongs*"(p. 94), whereas the i in *stile* "seems to be such a *Dipthong* [...] composed of a.i or e.i and not a simple Original Vowel" (p. 95). These observations are difficult to interpret. It could be taken to imply that he does not recognize /r/-vocalization as a "proper diphthong", but this is not made explicit in his writings.

5.4.3 Syllable Structure and Prosody

In conjunction with the concept of syllable, Holder comments on phonotactics. He states that there has to be "one Vowel in every Syllable, for varieties sake, sometimes preceding, sometimes following, and sometimes interposed between the Consonants" (p. 92). He fails to note, however, that while this holds true for English, it is by no means a language universal.

When it comes to what are today considered prosodic features, Holder distinguishes between "emphasis" (intensity and/or duration) and "accent" (pitch): "There some other Accidents besides those spoken of before, which have an Influence in varying the Sound of Languages, as *Accent* and *Emphasis*; which though now much confounded seem to have been formerly more distinguished. *Accent*, as in the *Greek* names and usage, seems to have regarded the Tune of the voice; the *Acute* accent raising the Voice in some certain Syllables to a higher, *i.e.* more acute Pitch or Tone, and the *Grave* depressing it lower, and both having some *Emphasis, i.e.* more vigorous pronunciation" (pp. 98-99).

He proceeds to discuss between-language differences in this respect between French and Ancient Greek. This, once again, is a very modern approach to comparative phonetics.

6   Holder on Speech Perception and Clinical Issues

In the "APPENDIX Concerning Persons DEAF AND DUMB", Holder touches upon the physiology of speech perception. He is well aware of the close interaction between speaking and hearing and of the fact that congenitally deaf individuals who do not receive treatment will not learn to speak: "[...] the Tong and the Ear, Speaking and Hearing, hold a correspondence, by which we learn to imitate the Sound of Speech, and understand the meaning of it" (p. 114). "[...] they who want that *Sence* of *Discipline* (*Hearing*) are also by consequence deprived of Speech [...]" (p. 115).

His understanding of the peripheral hearing mechanism seems somewhat limited. He regards a lax tympanic membrane as the principal cause of hearing loss/deafness. This in turn is caused by either a malfunction of the incus and malleus or laxness of the m. tensor tympani. "And I am of opinion, that the most frequent *cause* of *Deafness* is to be attributed to the Laxness of the *Tympanum*, when it has lost its Brace or Tension by some irregularity in the Figure of those Bones, or defect in that Muscle" (p. 113). His explanation for bone conduction is rather odd: "Now that which I would infer, is, That in *those* [i.e. deaf individuals, A.B.] generally the *Auditory Nerve* is sound, and by a *branch* of the same *Nerve*, that goes between the Ear and the Palat of the Mouth, they can make a shift to hear themselves, though their outward Ear be stopt by the Laxe Membrane to all Sounds, that come that way" (p. 129).

Holder evidently had intuitive knowledge of how to distinguish between peripheral and central causes of hearing loss: He asked the patient to hold a lute string between his teeth and, when he perceived sound originating from it, concluded that the auditory nerve was not damaged (p. 160).

The following remark points to Holder being well aware of the role of redundancy in speech perception: "Any *Equivocal* word spoken *alone*, cannot be determined to any one certain Sense and Signification by him that hears it; of which there are numerous examples in every Language: Yet the same word in *Connexion* of Speech, as part of a sentence, is understood as easily as any other" (p. 122).

7   Conclusion

At a time when awareness of the processes of speech production and speech perception were generally quite limited, individual scientists like William Holder display in their publications an amazing degree of "modern" thinking both with respect to general concepts like the source-filter model of speech production or the phoneme and the phonetic detail in the description of sound production. The degree of precision in the description of consonants as well as suprasegmentals well exceeds that of vowels. This contribution attempts to exemplify the import of his writings on general phonetics from today's perspective. The full scope of his findings – especially with respect to the teaching of deaf-mute patients – has yet to be analyzed in detail.

8   References

[1]   Holder, William (1669): *Elements of Speech. An Essay of Inquiry into the Natural Production of Letters: with An Appendix Concerning Persons Deaf & Dumb.* London.

[2]   Abercrombie, David (1948): Forgotten phoneticians. *Transactions of the Philological Society* 47, 1-34.

[3]   Rieber, R.W. / Wollock, Jeffrey L. (1975): William Holder on Phonetics and Deafness. An Introduction to the New Edition of Elements of Speech. AMS Press, New York, pp. i – xv.

[4]   Ohala, John (2004): Phonetics and Phonology then, and then, and now. In: Quene, H. / van Heuven, V. (eds.): On *Speech and Language. Studies for Sieb G. Nooteboom.* LOT Occasional Series 2, pp. 133-140.

[5]   Abercrombie, David (1993): William Holder and other 17th-Century Phoneticians. *Historiographia Linguistica* 20, pp. 309-330.

[6]   Laver, John (1978): The Concept of Articulatory Settings: An Historical Survey. *Historiographia Linguistica* 5, pp. 1-14.

[7]   Subbiondo, Joseph L. (1978): William Holder's 'Elements of Speech (1669)'. A study of applied English phonetics and speech therapy. *Lingua* 46, pp. 169-184.

[8]   Panconcelli-Calzia, Giulio (1943): *Leonardo als Phonetiker.* Hamburg: Hansischer Gildenverlag.

[9]   Kempelen, Wolfgang von (1791): *Mechanismus der menschlichen Sprache nebst Beschreibung einer sprechenden Maschine.* Wien. [Faksimile-Reprint ed. by Herbert E. Brekle and Wolfgang Wildgen. Stuttgart-Bad Cannstatt 1970].

[10] Amman, J.C. (1692): *Surdus Loquens.* Amsterdam: Wetstein.

[11] Bulwer, J. (1642): *Chirologia.* London: T. Harper.

[12] Wilkins, John (1668): An Essay Towards a Real Character And a Philosophical Language. London.

[13] Wallis, John (1674): *Grammatica Linguae Anglicanae.* Oxford: L. Lichfield.

[14] Firth, J.R. (1946): The English school of phonetics. *Transactions of the Philological Society,* 45, pp. 92-132.

[15] Panconcelli-Calzia (1961): *3000 Jahre Stimmforschung. Die Wiederkehr des Gleichen.* Hamburg: Hansischer Gildenverlag.

[16] Amman, J.C. (1700): *Dissertatio de loquela.* Amsterdam: J. Wolters.

[17] Cooper, Chr. (1687): *The English Teacher.* London: J. Richardson.

[18] Abercrombie, David (1967): *Elements of General Phonetics.* Edinburgh: Edinburgh University Press.

[19] Ladefoged, Peter (1971): *Preliminaries to Linguistic Phonetics.* Chicago: The University of Chicago Press.

[20] Catford, J.C. (1977): *Fundamental Problems in Phonetics.* Edinburgh: Edinburgh University Press.

[21] Abercrombie, David (1986): Hylomorphic Taxonomy and William Holder. *Journal of the International Phonetic Association* 16, pp. 4-7.

[22] Fant, Gunnar (1960): *Acoustic Theory of Speech Production.* The Hague: Mouton.

[23] Raphael, Lawrence / Borden, Gloria / Harris, Katherine ([6]2007): *Speech Science Primer. Physiology, Acoustics, and Perception of Speech.* Philadelphia etc.: Wolters Kluwer.

[24] Brooks, G.L. (³1965): *English Sound Changes.* Manchester: Manchester University Press.

[25] Lass, Roger (1987): *The Shape of English: Structure and History.* London: Dent.

[26] Viëtor, Wilhelm (1906): *A Shakespeare phonology, with a rime-index to the poems as a pronouncing vocabulary.* Marburg and London: Elwert and David Nutt.

[27] Wells, John (1982): *Accents of English. Vol.1.* Cambridge: Cambridge University Press.

[28] Jacobson, Roman / Fant, Gunnar / Halle, Morris (1952): Preliminaries to Speech Analysis. The Distinctive Features and their Correlates. Cambridge, Mass.: MIT Press.

Experimental phonetics at University College London before World War I

M. Ashby

Speech, Hearing and Phonetic Sciences, UCL (University College London)
m.ashby@ucl.ac.uk

Abstract: This paper argues that significant research and teaching in experimental phonetics at UCL preceded the establishment of the laboratory in 1912 by almost a decade. The forgotten work of E. R. Edwards [1] combined descriptive and experimental approaches in a pioneering study of Japanese. Recently recovered artefacts, and a close reading of the early publications of Daniel Jones (DJ), reveal a surprising range of equipment and techniques available before the laboratory was founded. In 1909, DJ's *Intonation curves* made use of what was arguably the first spoken corpus. It has been possible to locate and digitise the majority of the gramophone records of the corpus, and acoustic analysis confirms that DJ's f_0 representations are of astonishing accuracy. The view that the earliest phase of experimental phonetics teaching at UCL must have been derivative can be rebutted. On the contrary, it was research-driven, and almost certainly enlivened by demonstrations with relevant apparatus and informed by up-to-date scholarship.

1. Introduction

The techniques and instruments of experimental phonetics began to be settled from about 1890 onwards, particularly under the leadership of the abbé Rousselot (1846–1924) in Paris. Rousselot started to publish from 1891 [2], his *Principes de phonétique expérimentale* appeared in 1897 [3], and he set up a dedicated laboratory of experimental phonetics at the Collège de France in 1898. Similar laboratories soon followed elsewhere in Europe and around the world.

By 1916 one estimate [4: p. 62] put the number of phonetics laboratories worldwide at 'something more than twenty-five...most of which are in Europe'. The author does not give a list, but some idea of the proliferation of laboratories can be reconstructed from the retrospective notes in a later survey [5]. In France alone, further early laboratories were those established at Grenoble in 1904, under the direction of Théodore Rosset [6], and in the same year at Montpellier under Maurice Grammont [7].

The UCL laboratory did not start until 1912—and then with only a half-time assistant and very modest provision. By this time the laboratory at the Kolonialinstitut in Hamburg, under the direction of Giulio Panconcelli-Calzia (1878–1966) was establishing a reputation as the foremost in Europe. DJ and a nucleus of his early staff visited Hamburg for the Congress of Experimental Phonetics 19–22 April 1914 [8], and probably had the chance to examine the extensive facilities of the laboratory at first hand.

Published work from the UCL laboratory did not begin to appear until 1917, more than 25 years after Rousselot's revolutionary dissertation [2]. Overall, British efforts were certainly tardy, and it has been assumed that any teaching of experimental phonetics at UCL in the first decade of the twentieth century can only have been highly derivative [9: pp. 81–82].

Nevertheless, the present paper calls attention to evidence of significant experimental phonetics research and expertise at UCL from about 1903.

2. E. R. Edwards

In the years 1903–1905, Daniel Jones was preceded as phonetics lecturer at UCL by Ernest Richard Edwards (1871–1947). He was a teacher of modern languages at University College School from 1892; the young Daniel Jones was a student at the school 1897–1900, and the two certainly met at that early stage. Like many modern language teachers of the day, Edwards developed an enthusiastic interest in phonetics, and in 1899 he joined the IPA. He studied for a time at the University of Marburg with Wilhelm Viëtor, and from 1899 in Paris with Paul Passy, who encouraged Edwards to expand a student essay on the phonetics of Japanese into a major research project. Accordingly, in 1901–1902 Edwards undertook an extensive field trip to gather data throughout Japan, and in 1903 he completed and defended his doctorate on the phonetics of Japanese. It was published later the same year [1].

Alongside conventional phonetic description and transcription with IPA symbols, the thesis of Edwards was novel in making use of experimental data from palatography (using artificial palates) and kymography. Though largely forgotten today, it created a considerable stir on its appearance. Not only was it the first extended phonetic study of an East Asian language in its colloquial form, but the blend of descriptive and experimental approaches was entirely new.

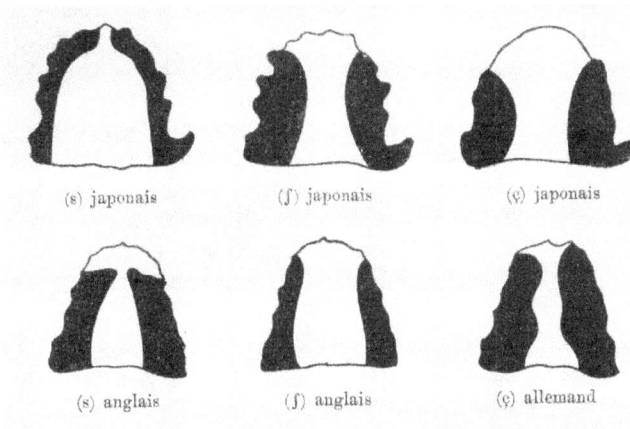

Figure 1: Palatograms from Edwards [1] showing the articulation of Japanese [s], [ʃ] and [ç] (top row) compared with corresponding sounds in English and German.

As an illustration of the data Edwards provides, Figure 1 shows his comparison of the articulation of sibilant fricatives from Japanese and English and of palatal fricatives in Japanese and German (p. 37), while Figure 2 shows a kymographic record of air pressure in an oral mouthpiece for a pair of words illustrating differences between single and geminate voiceless stops in Japanese (p. 29). He also states (p. 7) that a phonograph was used to settle difficult cases involving weak and voiceless vowels. No further details are given, but it seems possible that the device was used to facilitate auditory analysis by repeated or slowed-down replaying of certain portions of recordings.

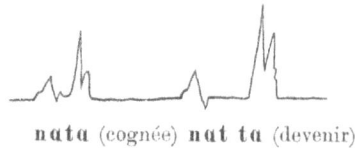

n a t a (cognée) **n a t t a** (devenir)

Figure 2: A kymogram from Edwards [1] illustrating a minimal contrast depending on consonant duration. The geminate [tt] of [natta] exhibits both a considerably longer hold phase than the single [t] of [nata] and greater airflow at release.

Edwards tells us that the kymograph he used was made in Paris, and in all probability it was the small portable model made by Charles Verdin (see Figure 3). The kymograms published by Edwards have only a single channel (mouth pressure) and evidently—at least in the set-up used by Edwards—the mouth-pressure tambour did not respond to vocal fold vibration, showing instead a smoothed indication of airflow even in voiced stretches of speech.

A paper for a non-specialist audience [10] published during Edwards's brief period as phonetics lecturer at UCL gives an insight into how he set about introducing the subject of phonetics to beginners, and suggests that his teaching probably used the same blend of descriptive and experimental approaches found in his thesis. In 1905, however, he left UCL to become an Inspector of Schools, giving up active work in phonetics. In 1907, the vacancy created by his departure was filled by Daniel Jones.

2. DJ's experimental phonetics teaching, 1909–1911

Though now remembered exclusively as an 'ear-phonetician', Daniel Jones introduced his own course 'Experimental Phonetics' which ran at UCL in the years 1909–10 and 1910–11, before the establishment of a laboratory. It consisted of 'A course of eight lectures illustrated with apparatus and lantern slides'. Collins and Mees [9: pp. 81–82] suggest that this course must have been rather derivative. They assume that the lantern slides must have illustrated work done by others, and don't comment at all on the 'apparatus' which the course description appears to imply.

In fact, we can find definite clues to what some of the slides may have contained, and very specific indications that DJ had at least some smaller and less expensive instruments at his disposal several years before the establishment of the laboratory. The first edition of DJ's *Outline of English Phonetics* [11] contains fairly extensive material on 'experimental' phonetics (the experimental material dwindles markedly in successive editions). The early descriptions are commonly accompanied by the names and addresses of makers and suppliers of equipment, and even by prices and the cost of postage.

For example, all the successive editions of the *Outline* retain his description of the fabrication and use of artificial palates, though with modifications to suit the passage of time. In the first and second editions there are instructions for making one's own palates, and the address in Paris of a Monsieur Montalbetti, who will make them to order (those made of gum-stiffened paper cost only 5 francs, metal or vulcanite are more expensive). By the third edition (1932), M. Montalbetti has disappeared, but a footnote still informs us that 'They can be made for about 10 shillings'. It seems very likely, then, that demonstrations with an artificial palate must have formed part of his 'Experimental' lectures.

Figure 3: Original print of a photograph used by DJ in the first edition of his *Outline of English Phonetics* (published in 1918, though largely written before WW I). The small clockwork kymograph is recognizable as the 'petit inscripteur' sold by the Parisian maker Charles Verdin. DJ is using a mouthpiece and a nasal olive, linked by rubber tubes to tambours of Marey pattern. The kymograph has a third tambour but no connection has been made to it. The smoking of the drum appears very uneven. Notice the pencilled instruction in German to the printer 'Seitenbreite (= page width) 11 cm'.

Late editions of the *Outline* also retain details of another instrument which had been mentioned from the beginning, a tube with an adjustable piston to demonstrate 'the effect of a resonance chamber in modifying quality of tone'. Even in the final (ninth) edition (1960) we are told that the makers are Messers Spindler and Hoyer of Göttingen. Earlier editions give the price (10 marks in 1922, 25 marks in 1932). DJ would hardly name the makers or specify the price unless he had handled a specific instrument, and probably purchased it. Besides, Paget tells us [12: p. 17] that DJ owned just such a device, and describes his own experiments with it, which must have been carried out in the 1920s.

Atkinson's Mouth Measurer appears in the first and second editions of the *Outline*, again complete with the maker's postal address and the price. The device was announced for sale in *Le Maitre Phonétique* in 1910. This was possibly a re-marketing at DJ's instigation, since it had first been developed almost 15 years earlier. Interestingly, DJ was in contact with both Henry Sweet and Bernard MacDonald around 1909–1910, and both of them, like DJ, provide testimonials to the usefulness of the device which appear on the printed instruction sheet found with surviving examples of the device.

Early photographs show Jones with a small clockwork kymograph (Figure 3). In numerous details it closely matches the 'petit enregistreur' sold by the French maker Charles Verdin [13: p. 86]. This surely implies that DJ had access to, and possibly owned, such an instrument before a much larger model was constructed for the UCL lab by the British firm of C. F.

Palmer. With a kymograph at his disposal, lantern slides prepared from original kymograms, or—since it was small and portable—demonstrations of the device itself, could therefore easily have formed part of DJ's lectures.

Nécessaire de Phonétique Expérimentale.

Par

Ad. Zünd-Burguet.

Diplôme d'Honneur, Exposition Universelle de Liège, 1905.   Tous droits réservés.

Le «Nécessaire de Phonétique expérimentale et pratique» contient un Cadran indicateur pliant à soufflet, à timbre mobile et à curseur (Modèle de Zünd-Burguet), un Signal du larynx à suspension élastique (Modèle de Z.-B), un Signal du larynx simplifié (Modèle de Z.-B.), une Embouchure en aluminium avec bouchon en caoutchouc et tube de verre (Modèle de Z.-B.), deux Ampoules en caoutchouc, une ronde et une plate, deux Olives nasales et environ un mètre de tube de caoutchouc.

Figure 4: The cover page of the instruction leaflet which accompanied Zünd-Burguet's experimental phonetics 'outfit' or 'kit' (French *nécessaire*). An artificial palate is visible in the open lid of the box, though no palatography materials were supplied with the kit and palatography is not mentioned in the leaflet. No other copy of the item has been located, and this example found in DJ's personal offprint collection may be a rare survival. DJ probably owned the kit itself, though none of the contents have been found.

In addition, his first work on the analysis of intonation [14] indicates that already by 1907 DJ must have had access to a good-quality gramophone. His account of Rosset's Grenoble laboratory [15] includes remarks on the superiority of disc recording over the cylinders used by Rosset, implying that DJ had evaluated the two sound recording methods critically.

DJ wrote in paragraph §87 of the *Outline*, 'The apparatus used in elementary instrumental phonetics includes the artificial palate, the kymograph, the laryngoscope, the mouth measurer, the gramophone and other talking machines, and a number of less important instruments.' As has been shown it seems likely that he would have been in a position to demonstrate most, if not all, of these in his classes between 1909 and 1911. The 'less important' instruments may have included his resonance tube, and the rather toy-like demonstration items [16: pp. 77–80] from an outfit sold by Zünd-Burguet (See Figure 4).

As for the lantern slides, there are relatively few indications of what they might have included, but we need not accept the assumption made by Collins & Mees that they were 'obviously an attempt to make up for the lack of a proper phonetics laboratory' [9: p. 82]. Undoubtedly some would have illustrated the work of others, but this is natural if the intention is to provide coverage of recently published research. Jones's reading was extensive, and his work for *Le Maitre Phonétique* kept him in contact with newly published books and journals. In his recently rediscovered offprint collection are two copies of a seminal paper by E. A. Meyer [17], and in one of these DJ has modified the caption of a figure, also adding pencilled instructions for a photographer on how to prepare slides from the page. The resulting slides, showing sectional diagrams of vowels and certain consonants from state-of-the-art X-rays, were probably intended for the 'Experimental' course. Among the lantern slides actually surviving at UCL, some showing pages from his own work *Intonation Curves* could have been made as early as 1909 and might well have found a place in his teaching in the period 1909–1911.

All in all, there is no reason to suppose that DJ's early 'Experimental phonetics' lectures were derivative and inferior. On the contrary, the indications are that the lectures may have combined practical demonstrations of basic concepts with accounts of the latest research findings, including his own—surely an excellent mixture.

3. *Intonation curves*

In 1909 DJ published a work with the full title *Intonation curves: A collection of phonetic texts, in which intonation is marked throughout by means of curved lines on a musical stave* [18]. It was based on the analysis of eight commercially available gramophone records. Three records were of English speech, three of French and two of German.

In addition to the intonation analysis, DJ gives (1) the orthographic versions of the texts, (2) a 'very detailed form of phonetic transcription', and (3) a transcription of the 'standard' pronunciation of each language. He was thus working with an ensemble of recordings and a set of time-aligned representations on several levels. It is not unreasonable to suggest, therefore, that DJ's 1909 material for *Intonation curves* constitutes an early—and possibly the first—spoken corpus.

The method of analysis was to listen repeatedly to the records, lifting the needle from the record at successive points and noting the 'the impression of the sound heard at the instant when the needle is lifted'. Pitch was assessed musically, with reference to a tuning fork as necessary. The resulting intonation curves are plotted on a musical staff, so as to align with the detailed segmental transcriptions, and divided into 'bars', each corresponding to one syllable of the text. The length of each syllable on the page is simply the width occupied by the corresponding section of the phonetic transcription, which is printed with normal character spacing (but no word spacing), and as a result the indicated lengths of the bars have little correspondence with real durations.

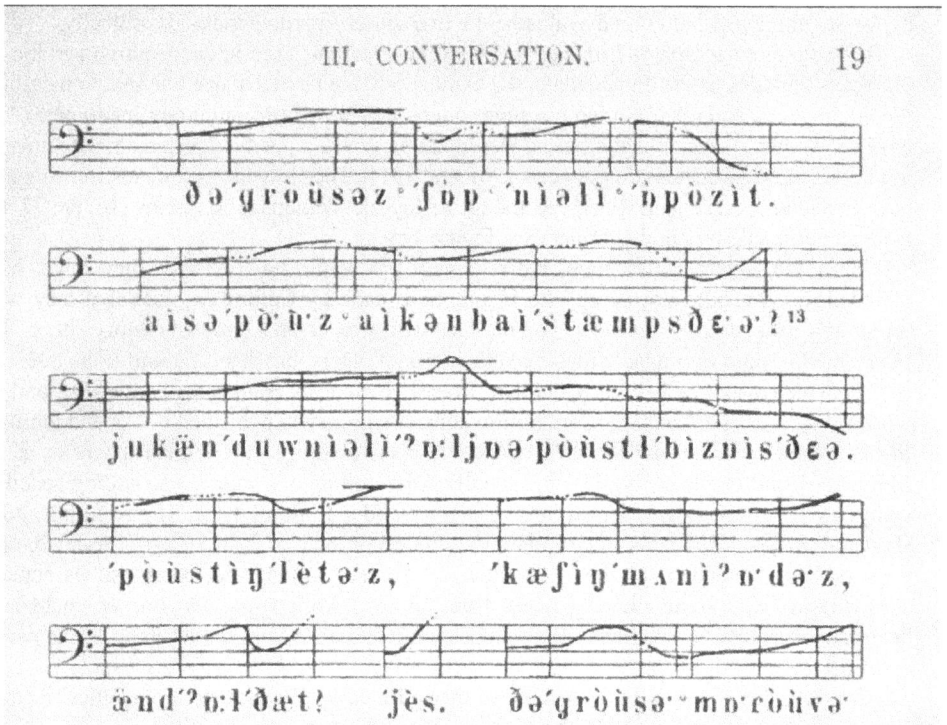

Figure 5: Part of a specimen page of the English conversation sample in *Intonation curves*, showing intonation notation aligned with the 'detailed' phonetic transcription. The corresponding orthographic and 'standard' phonetic versions are printed on the facing page.

DJ gives full particulars of the records he used, which has made it possible to seek surviving copies of them and attempt a partial verification of his findings. Six of the 8 have so far been located and digitized.

To convert DJ's musical representations into f_0 estimates, positions of successive points on the curves drawn on the staff must be estimated graphically. This was accomplished by viewing scanned images of the printed pages in high magnification and taking measurements with an on-screen cursor, yielding coordinates reckoned in pixels. These were entered into a spreadsheet which performs a mapping to corresponding log f_0 values by means of a lookup table (calculated on the basis of standard concert pitch, with A= 440 Hz).

Audio files were processed in SFS [19]. A fundamental frequency track was added, using 'fxrapt', an implementation of the RAPT algorithm [20], which was found to give very satisfactory results even with these old and relatively noisy recordings. Using a script to sample the f_0 at voicing onsets and offsets and other fairly conspicuous landmarks, it is possible to obtain corresponding pairs of f_0 determinations at sampling points throughout a phrase, each syllable being represented by 1, 2 or 3 points which can be time-aligned reasonably well.

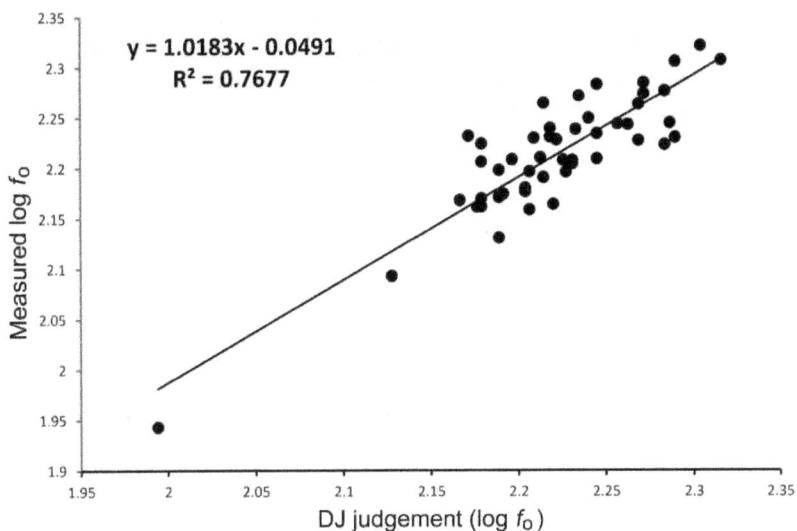

Figure 6: Correlation of 50 automatically-made estimates of f_0 with DJ's corresponding estimates, for a randomly-selected sample of French consisting of 25 syllables.

Scatterplots showing DJ's estimates versus measured f_0 can then be prepared. For the example shown in Figure 6, the correlation is high ($r^2 = 0.77$), indicating an excellent degree of agreement. The intercept of the fitted regression line is negative, tending to indicate that DJ's judgements are on the high side, and thus that he probably adjusted his clockwork gramophone so as to play this record a little faster than the modern standard of 78 rpm. Analysis of the other records shows a range of small discrepancies, both positive and negative. This was exactly as expected, since recording rates had not been entirely standardized at this time [21].

Much of the scatter in Figure 6 is probably owing to (unavoidable) misalignment in time of the selected sample points. An alternative approach is to present the modern f_0 estimates on a musical staff resembling DJ's. A further spreadsheet was devised which accepts f_0 values in Hz from a text file, and plots them on a logarithmic scale, along with constants (for the lines) and suitable graphic elements to complete the staff. A script was written which selectively stretches and compresses the time axis of the f_0 curve, so as to match the pseudo-time of DJ's bar lengths. The resulting plots are found to align with DJ's versions almost perfectly.

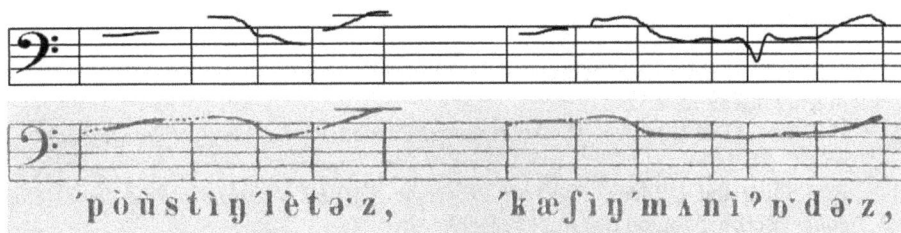

Figure 7: Top: automatically plotted intonation curve for a randomly-selected representative phrase of the English Conversation passage (male speaker); below: DJ's version.

Figure 7 is typical of the results. The two versions show a close agreement throughout. Notice for example the meticulously tracked pitch trough in the first syllable of *letters*, the small peak located in the second syllable of *cashing*, and how the greater part of the final rise is correctly placed in the second (unstressed) syllable of *orders*. Comparing the measured fundamental frequency with DJ's notated level in a suitable steady portion (for instance, the second syllable of *money*), it appears that DJ's judged pitch is about 4% flat—suggesting that he may have been listening to this particular record at approximately 75 rpm. DJ was personally acquainted with the speaker, Bernard P. MacDonald, so it is likely that the empirically selected rate he used gave values typical for the speaker.

Hart *et al.* [22] accord DJ's work on intonation 'dubious empirical status' and say 'impressionistic auditory descriptions remain difficult to interpret and may not be representative of other listeners' perceptions'. But as the present analysis demonstrates, DJ's judgements—though not presented in an obviously numerical form—are quantitative, precise, objective and verifiable. It is hard to see in what respect they fail to be 'empirical'. On the contrary, *Intonation curves* deserves to be recognized as a classic of scientific phonetics.

4. Conclusion

While it is true that the establishment of a phonetics laboratory at UCL came relatively late, it cannot be assumed that as a result the earliest phase of teaching and research at UCL lacked significant experimental content. Both the young DJ and his predecessor Edwards successfully blended 'taxonomic' and 'scientific' approaches [23] in interesting ways, and both produced work of striking originality which deserves to be better known and understood.

References

[1]   E. R. Edwards, "Étude phonétique de la langue japonaise," Impr. B.G. Teubner, Leipzig, 1903.

[2]   P.-J. Rousselot, "Les modifications phonétiques du language, étudiées dans le patois d'une famille de Cellefrouin (Charente)," Paris : Welter, 1891.

[3]   P.-J. Rousselot, *Principes de phonétique expérimentale.* Paris: Welter, 1897.

[4]   S. T. Barrows, "Experimental phonetics as an aid to the study of language," *The Pedagogical Seminary*, vol. 23, no. 1, pp. 63–75, 1916.

[5]   S. Pop, Ed., *Instituts de phonétique et archives phonographiques.* Louvain: Commission d'enquête linguistique, 1956.

[6]   L.-J. Boë and C.-E. Vilain, *Un siècle de phonétique expérimentale, fondation et éléments de développement: Hommage à Théodore Rosset et John Ohala.* Lyon: ENS, 2010.

[7]   J. Perrot, "Institut de phonétique de la faculté des lettres de l'université de Grenoble," in [5], pp. 229–232.

[8]   D. Jones, "The Congress of Experimental Phonetics," *Le Maître Phonétique*, vol. 29, pp. 50–51, 1914.

[9]   B. Collins and I. Mees, *The real Professor Higgins: The life and career of Daniel Jones.* Berlin: Mouton de Gruyter, 1999.

[10]   E. R. Edwards, "The phonetics of modern Japanese" [paper delivered 9 November 1904], *Transactions and Proceedings of the Japan Society, London*, vol. 7, pp. 2–26, 1907.

[11]   D. Jones, *An outline of English phonetics.* Leipzig und Berlin: B.G. Teubner, 1918.

[12]   R. A. S. Paget, *Human speech: Some observations, experiments, and conclusions as to the nature, origin, purpose and possible improvement of human speech.* London: K. Paul, Trench, Trubner & Co, 1930.

[13]   C. Verdin, *Catalogue des instruments de précision construits par Charles Verdin.* [Paris]: [J. Mersch.], 1890.

[14]   D. Jones, *Phonetic transcriptions of English prose.* Clarendon Press: Oxford, 1907.

[15]   D. Jones, "Phonetics at Grenoble," *Le Maître Phonétique*, vol. 24, pp. 143–146, 1909.

[16]   E. Galazzi, "Machines qui apprennent à parler, machines qui parlent: Un rêve technologique d'autrefois," *Études de Linguistique Appliquée*, vol. 90, pp. 73–84, 1993.

[17]   E. A. Meyer, "Röntgenographische Lautbilder," *Medizinisch-pädagogische Monatsschrift für die gesammte Sprachheilkunde*, vol. 17, pp. 225–243, 1907.

[18]   D. Jones, *Intonation curves. A collection of phonetic texts, in which intonation is marked throughout by means of curved lines on a musical stave.* Leipzig ; Berlin: B G Teubner, 1909.

[19]   M. Huckvale, *Speech Filing System.* London: UCL, 2013. http://www.phon.ucl.ac.uk/resource/sfs/

[20]   D. Talkin, "A robust algorithm for pitch tracking (RAPT)," in *Speech coding and synthesis*, W. B. Kleijn and K. K. Paliwal, Eds. New York: Elsevier, pp. 495–518, 1995.

[21]   T. Holmes, *The Routledge guide to music technology.* New York: London: Taylor & Francis, 2006.

[22]   J.'t Hart, R. Collier, and A. Cohen, *A perceptual study of intonation: An experimental-phonetic approach to speech melody.* Cambridge: Cambridge University Press, 1990.

[23]   J. J. Ohala, "There is no interface between phonology and phonetics: A personal view," *Journal of Phonetics*, vol. 18, pp. 153–171, 1990.

The "Bonn Connection" and its consequences: Paul Menzerath and Werner Meyer-Eppler's reunification of phonetics and phonology and the emergence of a new phonetic speech science based on Shannon's Mathematical Theory of Communication

Hans G. Tillmann, Jessica Siddins
LMU München, IPS

tillmann@phonetik.uni-muenchen.de
jessica@phonetik.uni-muenchen.de

Abstract: 1950, only two years after it was founded, the MIT's Electronic Research Laboratory (ERL) organised the first „Speech Communication Conference". Surprisingly, among the fourteen very prominent participants there were merely two representatives from the field of phonetic speech research: *Paul Menzerath* and *Werner Meyer-Eppler* from Germany; and it was on this very trip to the USA that they first met Claude Shannon of Bell Labs. Upon their return to the University of Bonn they founded the *Institut für Phonetik und Kommunikations-forschung*[1] which completely changed the direction of phonetics into speech communication research by forming a unique collaboration originally based on Shannon's *Mathematical Theory of Communication*. We begin with a discussion of how it took another eighty years following the great early successes in speech physiology and subsequent school of early instrumental phonetics from 1850 onwards for the classic phonetic theory of vowels and consonants to finally come crashing down. We then outline how this crisis was overcome by the paradigm shift at the institute in Bonn and describe the consequences for the subsequent development of phonetic theory and its applications. Finally, we demonstrate that – looking back – the eventual collapse of early instrumental phonetics was virtually pre-programmed and could be expected because of the initially successful but in hindsight far to optimistic "simplifying assumptions" made by speech physiologists such as Brücke (1849/1856) and Bell (1865).

N.B.: An extended version of this paper will be prepared for the Conference and made available on the internet for downloading.

1. Early modern phonetics since 1850 - from the initially great successes to the final disaster of instrumental work eighty years later

A full description of "Early Modern Phonetics, especially Experimental and Instrumental Work" has been given by the first author 1994 in EmP[2]. With respect to the paradigm shift described in section 2, it should be pointed out that, quite different from the situation in the US (and the UK), in the years after 1920 the academic development of speech sound research in central Europe had spiralled further and further into a deep crisis which finally led to

---

[1] It should be noted that to L1-speakers of German the term *Kommunikation* at that time was a strange foreign word and totally incomprehensible; it took a few years before e.g. the verb *kommunizieren* ('to communicate') became a familiar German loan-word.
[2] ENCOCLOPEDIA OF LANGUAGE AND LINGUISTICS (p. 3082 – 3095), reprinted in Koerner et. al (1995)

Trubetzkoy's strict partitioning of the scientific study of phonetics from purely linguistic phonology.

In the mid19[th] century, Brücke and with him the field of phonetic science split from the philological disciplines so that two different fields went their own ways. For Brücke "the time has come" to present the 'natural value' of speech sounds and to explain the 'natural connection between the sounds and their signs' (1856, p.1).

The *value* of any alphabetically given sound is determined by its properties which were called attributes, and Brücke distinguishes **two different ways** of finding these attributes:

> „Man kann die Art und Weise untersuchen, wie sie Nachbarlaute affizieren und von ihnen affiziert werden [...], um hieraus ihre Attribute herzuleiten. **Dies ist der Weg des Sprachforschers.**"[3]

> „Andererseits kann man directe Beobachtungen und Versuche über die Art und die Bedingungen ihrer Entstehung anstellen und hieraus eine Einsicht in ihre Natur und ihre Eigenschaften gewinnen. **Dies ist der Weg des Physiologen.**"[4]

Brücke (1849, 1856) in Austria and - only a few years later - Bell (1863, 1867) in England had learned[5] that certain articulatory configurations could be attributed to each vowel and consonant: for each vowel, there is a specific tongue height and position and lip aperture: rounded or unrounded. Other articulatory configurations are characteristic of consonants, in that a certain articulating organ may be associated with a certain place of articulation, and the degree of constriction determines the manner of articulation[6].

Those who referred to themselves as "sound physiologists" at that time were not true physiologists, but rather early behaviourists with one eye on the letters of the alphabet and the other on the visual or tactilely perceivable speech movements which could be looked at when spelling out the alphabetic sounds in isolation.

In view of the phonetic revolution in Bonn (and the development from articulatory phonetics to experimental and instrumental phonetics described in EmP), we can identify three important stages:

(1) Instrumental phonetics with its graphical representations of the 'speech curves' can be seen as an early form of speech signal processing. (We return to this point in Section 3).

(2) The interest of research underwent a major transformation. While the strict identity of articulatorily defined vowels and consonants as constant units were the deciding factor for Brücke, Bell and their successors, the new perspective was that the instrumentally measured time functions showed a large amount of variability in the execution of speech movements. It were 1891 the "modifications phonétiques du langage" in the title of Rousselot's dissertation that anchored experimentally reproducible systematic variability of speech sounds at the centre of scientific interest. A summary of the most important results from today's perspective can be found in Scripture (1902)[7].

(3) As stated in EmP: „The paradoxical situation was that instruments had been introduced to replace subjective hearing by objective measurements, but the resulting picture did not show

---

[3] One can look at the sounds (represented by letters) to see how they affect their neighboring sounds, and how they are themselves affected by these, [and one can also follow the changes they suffer during the course of time and through their going over from one language into another] in order to determine their attributes. **This is the way of the linguist**.

[4] A different possibility is that direct observations and experiments concerning the manner and condition of their formation lead to the desired attributes. **This is the way of the physiologist**.

[5] Mainly from observations made during educational instruction for the deaf; cf. EmP (p. 3084).

[6] Why is manner of articulation not designated in Jespersen's analphabetic system? Look for the answer in EmP.

[7] For example, it included VOT, which was re-discovered only much later, as well as pre-final lengthening and the algorithm for manually calculating FFT for the spectral analysis of a vocal period drawn with a pencil from a gramophone disk.

what was originally looked for: visible speech sounds. These had disappeared and had to be reinvented by the new phonologists of the Prague School Phonology." (p. 3092)

The situation around 1930 is well preserved in the following quote from Scripture (1932), who reports that the audience was shocked when the first cineradiographic film of running speech was shown at a meeting of the International Society of Experimental Phonetics organized by Menzerath 1930 in Bonn:

> „Die kleinsten Bewegungen der Lippen, der Zunge, des Gaumensegels, des Zungenbeins, der Kehlkopf-knorpel usw. spielen sich vor dem Auge ab". Scripture nennt den Eindruck eines solchen Films „über-wältigend." Denn da sehe man einen „Schattenmenschen […], wie er spricht, atmet und schluckt. Die Sprechwerkzeuge stehen nicht für einen Augenblick still, jeder Sprechakt ist die Summe der Bewegungen aller Organe des Mundes, des Rachens, des Kehlkopfes usw., und diese Summe spielt sich in der Zeit ab. Lautstellungen gibt es überhaupt nicht: es kommt alles auf Lautbewegungen hinaus. Man begreift sofort, daß die bisherige Lautphysiologie nur eine Irrlehre sein kann, und wartet gespannt auf neues."(S.. 173)[8]

Another prominent researcher, Panconcelli-Calzia, argued that sound segments or syllables were not a phonetic reality at all. He considered such units pure fiction and an invention of linguists.

2. The Bonn Connection's way out of the cul-de-sac – a paradigm shift in phonetic speech science

If - according to Eli Fischer-Jörgensen - "all phoneticians are something else" (p.c.), then this clearly applies to the two Protagonists that caused such an innovative change of phonetic speech research.

Paul Menzerath (1883-1954), a Professor of Psychology at the University of Gent, left his chair when the University of Bonn offered him the possibility to work in the field of experimental phonetics. He installed a small Lab for instrumental and experimental phonetics which since 1928 published a whole series of excellent results (cf. EmP p. 3091 ff). For our context here the 1933 study on "Koartikulation, Steuerung und Lautabgrenzung" is of outstanding importance. It opened the door for a new phonetic theory of speech sounds in terms of vowels and consonants and explained their specific role in the articulatory production of syllabically well-formed natural speech utterances. When Menzerath and Lacerda discovered the phenomenon of coarticulation they were able to propose a *phonetic* way out of the dilemma described above. They used the kymographic method to show that the speech sounds at the beginning of the syllable are coarticulated, i.e., produced at the same time, and that the vowel in a syllable rhyme is stopped ('controlled') by the following consonant if there is one. Menzerath and de Lacerda observed various coarticulatory pheno-mena; however, in modern terms, *Koartikulation* for them mainly seems to refer to the anticipation of consonantal and vocalic features in syllabic initial position, whereas carry-over effects cannot easily be related to their concept of *Steuerung*. They formulated the new theory that the complex combination of simultaneous movements of the speech organs, which they called 'Synkinese', serves only for the purpose of producing 'acoustically' a clear sequence of separated and therefore directly segmentable sound units.[9] In any case they were able to

---

[8] The impression of such a film is overwhelming. The organs of speech do not remain still for an instant, every speech act is the combination of movements of all organs of the mouth, the throat, the larynx etc., and this combination is deployed over time. Sound positions simply do not exist. One understands at once that the sound physiology up to now has been based on an illusion and one awaits new explanations.

[9] In EmP it was pointed out, that Menzerath and Lacerda "failed to make any distinctions between purely acoustic sound segments in the physical world and the auditorily perceived sounds which again were taken as a common reality, an assumption that would later lead to a repetition of the first crisis" (p. 3093)

answer the open questions of "Wohlartikuliertheit"[10] that Techmer had asked in his inaugural lecture 1871 in Leipzig: "Naturwissenschaftliche Analyse und Synthese der hörbaren Sprache". An elaborated version was published in Techmer (1874); more about this is presented in EmP (p. 3087).

Werner Meyer-Eppler (1913-1960) studied mathematics, physics, and chemistry, and when he had presented his postdoctoral thesis on "Periodenforschung" (the title of his *Habilitationsschrift*[11]) he proved to be the world's leading expert in applying new mathematical and statistical methods to electronic speech signal processing.

It was an incredible stroke of luck that in the years after World War II the world's leading specialist of experimental phonetics (with his new acoustically based theory of alphabetically referable units of speech in a given language) came together with the much younger physicist Meyer-Eppler, the leading specialist in the field of acoustic speech signal production and transmission. Inviting him to become a researching member of his laboratory was the very beginning of the paradigmatic shift which in the first author's contribution to the Heike-Festschrift has been called "Bonner Wende" (much more details in Heike-FS (2013)). Already in 1949 Meyer-Eppler's published his very influential monograph "*Elektrische Klangerzeugung. Elektronische Musik und synthetische Sprache*"[12]. And it was due to Meyer-Eppler's initiative and his connections to Bell Labs and to the MIT that they both attended the first Conference on Speech Communication as mentioned in the Abstract. After their return they founded the new *Institut für Phonetik und Kommunikationsforschung* at the University of Bonn (belonging to the philosophical faculty of this University).

Four years after its formation, from 1954 to 1960, Meyer-Eppler took over the sole leadership of the institute. During these years he carried out a whole set of far-reaching activities[13]. The two most important ones must be mentioned here explicitly[14].

The first most important scientific activity of Meyer-Eppler was to develop - on the bases of Shannon's Mathematical Theory of Communication - his own completely new version of *Information Theory* which then appeared 1959 as the first Volume in the Series „**Kommunikation und Kybernetik in Einzeldarstellungen**" (edited by himself).

The second type of activity consisted in creating a whole set of research projects which were all financed externally. So Meyer-Eppler was able to offer research positions to a group of highly qualified young scientists which he attracted to Bonn.

Georg Heike was interested in Electronic Music[15], but Meyer-Eppler convinced him to investigate the acoustic features of the phonemes of the dialect of the city of Cologne.

Gerold Ungeheuer came from a technical university with degrees in electronics and communication engineering. In his dissertation supervised by Meyer-Eppler he used physics to solve the problem of computing the resonance frequencies of the vocal tract given the area-

[10] In section 3 below we will return to the concept of „well-articulatedness" which has been extensively dealt with in TmM (1980) the first author's book on "Phonetics: Spoken Signs, Speech Signals, and Speech Communication.

[11] By which he received 1942 the qualification as a university lecturer in physics at the *Faculty for Mathematics and Natural Sciences* of the Bonn University.

[12] Electric sound-generation: Electronic Music and Synthetic Speech

[13] In the more extensive version of this paper we will present a summary of the many diverse research projects initiated and conducted by Meyer-Eppler during the six years until his early death.

[14] More details are given in Heike-FS including the fact that Meyer-Eppler caused Manfred Schröder's move from the 3rd Physical Institute in Hanover to Bell Labs in the US (pc).

[15] In Elena Ungeheuer (1992) the complete story is given about the role, Meyer-Eppler played in the creation of the "Studio für Elektronische Musik" at the broadcastingstation "Kölner Rundfunk". The leading composer of this studio, Karlheinz Stockhausen, became a PhD-student of Meyer-Eppler and visited his lectures on information theory for several semesters.

function of the vocal tract, which determines the formant frequencies of the respective vowel. In 1957, he had already discovered what later has been called the *quantal nature of acoustic speech production*, but this has been widely ignored up until now in the English written literature (Jim Flanagan is one of the exceptions[16]).

Helmut Schnelle was a nuclear physicist and became a specialist of symbolic logic, artificial intelligence, and computer linguistics. He started the DAWID-Project with Heike and Rupprath before he left the group in Bonn to take a chair in Berlin.

The "Bonn-Connection's" idea of reunifying instrumental phonetics and phonology can be very easily depicted by Meyer-Eppler's schematic presentation of the Verbal Communication Chain. Fig. 1 is taken from page 2 of his Book "Grundlagen und Anwendungen der Informationstheorie" (Principles and Applications of Information Theory):

Die sprachliche Kommunikationskette

Sprachliche[2] Kommunikation setzt als Expedienten in der Regel ein menschliches Individuum voraus. Das von ihm dem Perzipienten übermittelte Signal ist als Träger sprachlicher Funktionen *Zeichen* kraft seiner Zuordnung zu geistig erfaßten Gegenständen und Sachverhalten[3]. Die Zuordnung selbst ist *beliebig*[4] und Ergebnis einer *Setzung* oder einer besonderen *Übereinkunft* zwischen dem Expedienten und dem Perzipienten; der Zeichencharakter wird dem Signal *verliehen*, er haftet ihm nicht an wie das Symptom.

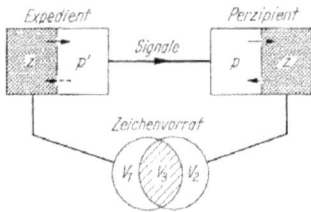

Abb. 1; 3. Modell der einfachsten sprachlichen Kommunikationskette. V_1 aktiver Zeichenvorrat des Expedienten, V_2 passiver Zeichenvorrat des Perzipienten, V_3 gemeinsamer Zeichenvorrat

Die einfachste *sprachliche* Kommunikationskette (Abb. 1; 3) weist also, im Gegensatz zur Beobachtungskette und diagnostischen Kette, eine *doppelte* Verbindung zwischen den beiden Kommunikationspartnern auf. Neben der realen, mit physikalischen Methoden nachweisbaren

[2] Das Wort *Sprache* (language) wird hier in dem allgemeinen Sinn verwendet, den ihm unter anderen CH. W. MORRIS beilegt (Foundations of the theory of signs. In: International encyclopedia of unified science, Bd. 1, S. 77–137. Chicago: Univ. of Chicago Press 1955).

[3] BÜHLER Sprth S. 28 ff.

[4] DE SAUSSURE GSpr S. 79 ff.

Fig. 1

The following statement is a citation from the editor's foreword:

„Aufgabe der Informationstheorie ist es, die Kommunikation von Mensch zu Mensch, die sich als Zeichenverkehr manifestiert, oder die Kommunikation des Menschen mit der Welt, die auf eine Beobachtung hinausläuft, einer quantitativen und strukturellen Erfassung zugänglich zu machen, während die Kybernetik als „science of relations" (N.WIENER) die regulären Verhaltensweisen von hochkomplexen energetisierten „Systemen" (d.h. von informationsverarbeitenden „Maschinen", Lebewesen und Gruppen von Lebewesen) mit mathematischen Methoden studiert" (ohne Seitenzahl).[17]

[16] His wonderful book on "Speech Analysis, Synthesis, and Perception" appeared 1965 in Meyer-Eppler's Series mentioned above; cf. also Schroeder 1967.

[17] Information theory aims to make different types of communication quantitatively and structurally tangible. Communication of humans with each other is manifested in the exchange of signs, while the communication of

This quotation shows the broad communication-theoretical framework in which experimental phonetics and phonology come together. In particular, Trubetzkoy's distinction between relevant and irrelevant features played an important role in Meyer-Eppler's thinking (and teaching) and received a quite new interpretation.

Methodologically, Brücke's concept of 'direct observation' was replaced by the newly established 'authority' of an *external observer* ("externer Beobachter"):

„Die in einer Kommunikationskette sich abspielenden Prozesse können nur von einem außerhalb der Kette stehenden *externen Beobachter* hinreichend exakt beschrieben werden, einem Beobachter, dem *sämtliche* Glieder der Kette zugänglich sind. Zur Beschreibung des Beobachteten bedient er sich einer wissenschaftlichen *Metasprache*, die nicht mit der zwischen dem Expedienten und Perzipienten vereinbarten *Objektsprache* übereinstimmt. Alle informationstheoretischen Ausführungen der folgenden Kapitel sind in der Metasprache des externen Beobachters formuliert" (Meyer-Eppler 1959, S.5f)[18].

We cannot deal here in detail with the very complex relations between the "gemeinsamer Symbolvorrat", which is to be categorically identified on all linguistically relevant levels on the one hand and, on the other hand, the production and reception of the phonetically produced utterance with which a speaker addresses himself to a listener. In Kohler-FS (1993) the first author has discussed these problems by asking the question which *phonetic facts* a speaker must create when conducting an act of speech.

There is another point of interest which can be inferred from Fig.1. Whereas the traditional 'narrower' Type of phonetics has clearly always been a small subpart of linguistics, in Meyer-Eppler's schema of the verbal communication chain linguistics becomes only a small, but necessary subpart of phonetic speech science. It is needed in order to be able to specify the *shared symbol inventory* ("Gemeinsamer Symbolvorrat").[19]

Following his untimely death at the age of 47, his work could be continued because of his ongoing, successfully running research projects. This is the main reason why Meyer-Eppler's research group stayed active and continued to grow. Ungeheuer, Schnelle, Heike and others not only continued their work in the following years, but also began working in relatively new areas. The faculty invited guest professors. Max Mangold came from Saarbrücken to brilliantly teach narrow phonetic transcription. Göran Hammarström gave an introduction to the traditions of diachronic and synchronic linguistics and explained his system of linguistic units which was also published in Meyer-Eppler's series as volume V (then edited by the computer scientist Steinbuch who later introduced the term "Informatik", the new German word for Computer Science).

In 1963, Ungeheuer became officially responsible at the institute for „Kommunikations-forschung"[20]. Under his guidance, the electronics lab established by Meyer-Eppler and run by

humans with the world can be reduced to observation. Instead, as the "science of relations" (N. WIENER), cybernetics is the study of regular behavioural patterns found in highly complex energetised "systems" (such as "machines" designed to process information; living creatures and groups of those) using mathematical methods.

[18] The processes taking place in a communication chain can only be described in sufficient detail by an *external observer* who is able to access *all* the links in the chain. In order to describe what he observes he makes use of a scientific *meta-language* which does not correspond to the *object language* agreed upon by the sender ("Expedient") and the receiver ("Perzipient"). Thus, all information-theoretic explanations given in the following chapters are formulated in the meta-language of the external observer.

[19] This can also be inferred from the three footnotes of page 2 (in Fig. 1).

[20] He succeeded in getting Klaus Kohler from Edinburgh who confronted his new colleagues in Bonn with the English tradition of careful phonetic ear training; Helmut Richter came from Zwirner's institute and discussed with the present author the psychological aspects of speech communication; the linguist and specialist in foreign language teaching Franciszek Grucza stayed as a guest scientist from Poland in the institute until he received a leading position at the University of Warsaw (together with a professorship in language teaching).

R. Rupprath developed the first well functioning speech recognition system, the **Device for Automatic Word Identification through Discrimination, DAWID**. It was restricted to identify twenty Italian words, the numbers from "zero" to "nove" and command words such as "meno", "piu", "per", "diviso", "uguale", "dacapo" etc.[21]

The presentation of this great success in a small paper (by Tillmann, Heike, Schnelle, and Ungeheuer) at the International Acoustic Conference 1965 in Liège caused BMVtdg (the defence department of the German Government) to contact Schnelle and Ungeheuer offering them more than enough money to start a broad ensemble of application oriented phonetic speech research projects with the final aim of reliable automatic speech recognition of spoken German, but also projects were financed on Speaker Identification and Speaker Verification (cf. Tillmann 1973). This enabled Ungeheuer a tremendous extension of the number of his research projects. In the framework of this application-oriented work financed by the German Government, Meyer-Eppler's early *electronic* speech signal processing was, within a short time, replaced by the new much more powerful methods of *digital* signal processing. Thus the computer became the single most important instrument in phonetic speech research in Germany.

Soon after Bonn this was the case also in Köln, in Munich, and in Kiel.

3. Looking back on the first century of academic phonetic speech research

Phonetics as speech science in the sense of Menzerath and Meyer-Eppler has become an integral part of the large and complex research field of SLP (Spoken Language Processing)[22]. It therefore makes sense to consider the development of early modern phonetics as outlined above in Section 1 from today's SLP perspective[23]. In the following we consider a set of eight critical points which have been ignored during the first one hundred years of early modern phonetics.

(1) Early phoneticians presumed that in all spoken languages of the world there is no natural oral speech act without an actual utterance produced by a real speaker who utilizes his speech organs according to the pronunciation rules of his language, dialect, and sociolect. However, the simplified view that in an oral speech act one only has to distinguish between *directly observable utterance* and the *verbal meaning* semantically associated with it mentally was (as we will show) wrong[24]. This becomes evident as soon as we consider that the term 'utterance' is ambiguous at least in two different respects. While it was totally sufficient for linguists such as Bloomfield or Trubetzkoy to assume that "recurrent utterances are alike or partly alike", phonetic speech scientists were responsible for investigating the utterances produced by a speaker as phonetically given facts in order to find out why and how, in a given speech community, recurrent syllables, consonants, vowels, diphthongs etc are alike or different. However, there are two methodological mistakes which we would like to describe and clarify in (2) and (3) below.

[21] More detailed information in Heike-FS (p. 11ff).

[22] This widely accepted term was proposed by Hiroya Fujisaki in direct contrast to NLP (Natural Language Processing), which according to Adrian Fourcin could also be an acronym for "non-spoken language processsing" (pc).

[23] We will not comment on Techmer's thorough and comprehensive approach of a "naturwissenschaftliche Analyse und Synthese der hörbaren Sprache" from today's perspective, as we plan to discuss this in a further paper.

[24] We also find this misleading assumption in Bloomfield's famous definition D1: "An act of speech is an utterance".

(2) First of all, speech acts cannot be separated into isolated phonetic and semantic parts[25]. In reality, the concept of an utterance is ambiguous in a way that has been dealt with extensively and made made clear in TmM: each time a speech act occurs, the external observer registers the material processes taking place throughout the whole communication chain from brain to brain. These material processes can be measured in terms of analogue time functions $f(t)$ which can be digitalised and further processed using the methods of SLP. This type of data remains absolutely transphenomenal for all subjects involved in the speech act, unless there is a microphone and an oscillograph available.

(3) In order to understand how it is possible to establish whether two utterances are alike or partly alike from a categorical perspective it is necessary to distinguish *between two types of speech acts*. They differ *depending on the **intention** of the speaker*. In the first case, the speaker produces an utterance (e.g. "xyz") in order to confer the meaning xyz. The speaker and the listener transcend the perceived utterance mentally into the concrete or abstract situations in their worlds, which are semantically governed according to the linguistic structure of "xyz".

In the second case, *the speaker's intention* is to *mean the utterance itself*. This is the case when, for example, Daniel Jones speaks a series of isolated speech sounds into a microphone to record onto audiotape (in multiple *categorically identical reproductions*) in order to demonstrate the category of his cardinal vowels. The IPA vowel symbols are thus ostensively defined by the presentation of these clearly perceivable citation forms.

From a logical viewpoint, the utterance is intended to be **autonymic** (meaning itself). On the other hand, in the first case, we are dealing semantically with a **heteronymic** utterance[26]. In everyday speech autonymically and heteronymically produced utterances differ not only in their semantics, but also in their phonetic forms[27]. Normally, we are immediately able to detect with which of the two intentions a speaker produces his speech utterance.

(4) The term "utterance" is also ambiguous in a second way. We need to distinguish between the time functions registered by the external observer and the symbolic representation of the given sound categories. These are two fundamentally different empiricisms that are logically independent of each other, i.e. contingent (Feigl 1958). The relationship between the two types of speech data is purely empirical and can therefore - as in the case of categorical perception - be reproduced experimentally. Early phoneticians overlooked the difference between logical identity, which is expressed with the equal sign, and empirical identification which must be tested. They were naïve realists. Today, as critical realists, we know that the empirically given correlations between signals and symbols, i.e. time functions and categories, must be revealed by systematic research and have to be made explicit by empirically verifiable relating theories. It is thus no wonder that early instrumental phonetics with its naïve aims was doomed to fail.

---

[25] This mistake is also made by philosophical speech act theorists, who ask and try to establish how it can be possible that a speaker produces an utterance, "xyz", in order to express meaning xyz. Here the utterance of the speaker is given in quotation marks as if it could exist by itself. The answer that is given is that the semantic meaning of xyz (without quotation marks) arises out of the ***speaker's intention***, which the listeners must recognise in order to understand the speaker. We return to this in the next footnote.

[26] Here language philosophists should ask the question how it can be possible that a speaker produces "xyz" in order he means "xyz" as such itself. As the example of the cardinal vowels shows, it is really the case that the listener understands that Daniel Jones has the intention of demonstrating the phonetic category of a certain cardinal vowel by repeating categorically identical repetitions of that certain vowel.

[27] This can be demonstrated by comparing the words from spontaneous speech with their citation forms.

(5) Today, we can view the early phoneticians as naïve realists in yet another light. They were not aware of the importance of distinguishing between the bottom-up and top-down processes which always play a role during the cognition of autonymically or heteronymically produced utterances. By ignoring top-down processes the early phoneticians believed they were able to directly observe in a bottom-up way the alphabetic form of orally produced utterances in the physical reality. Today, we know that the listener's cognitive system needs to use top-down processes in order to interpret information transmitted in the acoustic signal by empirically verifying the L1 categories acquired during child language acquisition. The same is true for any phonetically acquired L2 knowledge.

(6) With the arrival of the DAWID project, the word as a category in the mental lexicon (Meyer-Eppler's "Symbolvorrat") suddenly became a central phonetic unit[28], because lexical units play a crucial role not only in the listener's perception of a natural speech act, but also in automatic speech recognition. A very good example of this is the **M**unich **Au**tomatic **S**egmentation tool, **MAUS**, developed at the IPS[29].

(7) It is only possible to investigate the actual realisations of the phonetic form of the words of an utterance (compared with their citation forms) if we know the canonical form of the given utterance. For example, in addition to the acoustic signal of an utterance, Munich MAUS requires a canonical form or an orthographic transliteration plus the name of the language in order to produce a fine phonetic transcription and acoustic segmentation.

(8) Our example of MAUS-annotated sound segments can be used to demonstrate our final point. This relates to the fact, neglected by the early speech physiologists, that natural speech articulation is typically produced far too quickly to be processed adequately by bottom-up processes, as coarticulated CvCvC sequences (C standing for consonants, including none, v for a vowel or diphthong), are characterised by their C prosody (see TmM's distinction between slow A, rhythmic B and fast C prosody (p. 39 ff))[30].

Brücke treated the letters of the alphabet as isolated lexical units and subjected them to direct observation in their citation forms. This explains how he came up with the absurd idea of improving the orthography by adapting it to actual pronunciation. The actual phonetic notation of such real pronunciations, such as those computed by MAUS, cannot possibly be used to deliver a useful orthographic representation. The cognitive systems of speakers and listeners have learned to process what from a lexical point of view could well be called "modifications phonétiques du langage". In order to develop empirically verifiable theories explicating the functioning of all the processes that must run effectively in an act of successful speech communication in the interdisciplinary field of SLP it becomes necessary (in the words of another title of Rousselot) to use more than one "méthode graphique appliqué à la recherche des transformations inconscientes du langage".

Indeed, Francis Nolan and the first author, both quite well trained in narrow phonetic transcription, had no idea what their tongues where doing when producing the combined EPG and EMMA data in fast spoken German or English, respectively, for Barbara Kühnert's PhD thesis on alveolar-velar [t-k] assimilation (Kühnert 1994). Thus, they were both quite surprised when they saw the very complex systematic picture presented by their student after she

---

[28] Cf. Kohler (1998) and Tillmann (1998)'s contributions to the conference "The word as a phonetic unit" organised by Pompino-Marschall in Berlin.

[29] https://clarin.phonetik.uni-muenchen.de/BASWebServices/#/services

[30] Cf. Phil Hoole's demonstration of A, B and C prosody:
http://www.phonetik.uni-muenchen.de/~hoole/kurse/artikul/abc.pdf

had looked through all the data as a scientific external observer in order to analyse the tongue behaviour[31] of the German and English subjects.

PS:

The idea to write this paper was born two years ago in Venice when the first author had given, invited by Jonathan Harrington, one of his penultimate looking-back talks at VIU, and in the audience the second author proposed to compare the early results (such as intrinsic pitch of vowels, the P-center, or those other ones mentioned above in footnote 7) with the results of today's research in more detail. As one can see this idea is still reflected in our section 3 above.

Here, the first author would like to add that he is going to prepare a fourth section of the paper (to be presented in Dresden) in order to discuss the question which role some really innovative ideas in the recent past of phonetic speech research could or should play in future SLP work.

He believes that there are at least four important topics that should not be neglected in future *application oriented* SLP-research:

(i)   Is the proposal to develop a **CPT** (*Complete Phonetic Theory*) of spoken German[32] - making use of the huge available databases collected in Verbmobil and BAS[33] - definitely out of date?

(ii)   Why has Hartmut Pfitzinger's amazing success in measuring the variation of local speech tempo and to depict it graphically like a smoothed F0-contour[34] been more or less ignored in SLP (up to now)?

(iii)   What is the future of speech *synthesis-by-analysis*[35]?

(iv)   How could in the near future phonetic speech signal processing contribute to improve the pronunciation of L2-speakers in Chinese[36], in German[37] or other foreign languages[38]?

---

[31] While in the EPG-Data there **is** either an alveolar [t]-contact **or is not** before the alveolar contact of the following [k]-word, the EMMA-date show all degrees of tongue tip movements from zero to full contact (as in the [k-k] control items).

[32] Tillmann, H.G. & B. Pompino-Marschall: Theoretical Principles Concerning Segmentation, Labelling, and Levels of Categorical Annotation for Spoken Language Database Systems, EUROSPEECH 1993, pp. 1691ff.

[33] Hess, W., Kohler, K., Tillmann, H. G. "The PhonDat-Verbmobil Speech Corpus", Proceedings of EURO-SPEECH, pp. 863-866, Madrid 1995

[34] Pfitzinger, H.R : Phonetische Analyse der Sprechgeschwindigkeit. , FIPKM 38, pp. 117-264, 2001. (This outstanding milestone of phonetic speech research should be printed as a book and also translated into English.) (one can download it from http://www.ipds.uni-kiel.de/hpt/)

[35] Tillmann, H. G.; Pfitzinger, H. R.: Parametric High Definition (PHD) Speech Synthesis-by-Analysis. Proc. ICSLP 2000, vol. III, pp. 295-297. Beijing 2000

[36] Tillmann, H.G.; Pfitzinger, H.R. (2004): Applying the Munich Parametric High Definition (PHD) Speech Synthesis System to the Problem of Teaching Chinese Tones to L1-Speakers of German. Proc. of the Int. Symposium on Tonal Aspects of Languages: Emphasis on Tone Languages (TAL), pp. 185-188. Beijing 2004

[37] Bissiri, M.P.; Pfitzinger, H.R.; Tillmann, H.G. (2006) Lexical Stress Training of German Compounds for Italian Speakers by means of Resynthesis and Emphasis. Proc. of the 11th Australasian Int. Conf. on Speech Science and Technology (SST 2006), pp. 24-29, 2006

[38] Tillmann, H.G.; Pfitzinger, H.R. (2004): The Development of an Advanced SLP-based System for the Individual Learning and Fast Training of Speaking Skills in a New Foreign Language. Proc. of the InSTIL/ICALL2004 Symposium on Computer Assisted Language Learning, pp. 17-20. Venice 2004

References

Bell, A. M.: Visible Speech. Universal alphabetics or self-interpreting physiological letters for the writing of all languages in one alphabet. London/New York 1867

Bissiri, M.P.; Pfitzinger, H.R.; Tillmann, H.G. (2006) Lexical Stress Training of German Compounds for Italian Speakers by means of Resynthesis and Emphasis. Proc. of the 11th Australasian Int. Conf. on Speech Science and Technology (SST 2006), pp. 24-29, 2006

Brücke, E. W. v.: Untersuchungen über die Lautbildung und das natürliche System der Sprache. Sitzungsber. der königl. Akad. der Wissenschaften. Mathem.-Naturwiss. Classe II, 182-208, Wien 1849

Brücke, E. W. v.: Grundzüge der Physiologie und Systematik der Sprachlaute für Linguisten und Taubstummenlehrer. Wien !956

Feigl, H.: The „"Mental" and the „"Physical", in Volume II of *Minnesota Studies in the Philosophy of Science: Concepts, Theories, and the Mind-Body Problem, edited by Herbert Feigl, Michael Scriven, and Grover Maxwell*, Univ. Minnes. Press, 1958.

Hammarström, G.: Linguistische Einheiten im Rahmen der modernen Sprachwissenschaft. Berlin-Göttingen-Heidelberg 1966

Heike, G.: Sprachliche Kommunikation und linguistische Analyse. Heidelberg 1969

Hess, W., Kohler, K., Tillmann, H. G. "The PhonDat-Verbmobil Speech Corpus", Proceedings of EUROSPEECH, pp. 863-866, Madrid 1995

Kemp, J. Alan. (1994). Phonetic transcription: History. In R. E. Asher & J. M. Y. Simpson (Eds.), *The encyclopedia of language and linguistics* (Vol. 6, pp. 3040–3051). Oxford: Pergamon.

Koerner, E. F. K. and R. E. Asher (Eds.): Concise History of the Language Sciences from the Sumerians to theCognitivists. Cambridge UP, 1965

Kohler, K. J.: The disappearance of words in connected speech. ZAS Papers in Linguistics 11, 21-34, Berlin 1998

Kühnert, B. : Die alveolar-velare Assimilation bei Sprechern des Deutschen und des Englischen - kinematische und perzeptive Grundlagen. München 1964

Menzerath, Paul und A. de Lacerda: Koartikulation, Steuerung und Lautabgrenzung. Berlin-Bonn 1933

Meyer, E. A.: Beiträge zur deutschen Metrik. Die neueren Sprachen 6, 1-37, 122-40, 1897

Meyer-Eppler, W.: Elektrische Klangerzeugung: Elektronische Musik und synthetische Sprache. Bonn 1949

Meyer-Eppler, W.: Grundlagen und Anwendungen der Informationstheorie. Berlin-Göttingen-Heidelberg, 1959

Meyer-Eppler, W. und G. Ungeheuer: Die Vokalartikulation als Eigenwertproblem. Zeitschr. für Phonetik 10, 245-257, 1957

Pfitzinger, H. R : Phonetische Analyse der Sprechgeschwindigkeit. FIPKM 38, pp. 117-264, München 2001

Pompino-Marschall, B.: Einführung in die Phonetik. Berlin 2003

Rousselot, P.J.: Les modifications phonétiques du langage. Revue des patois gallo-romans 4, 65-208, 1981a

Rousselot, P.J.: La méthode graphique appliqué à la recherche des transformations inconscientes du langage. Revue des patois gallo-romans 4, 209-13,1981b

Schroeder, M. R.: Determination of the geometry of the human vocal tract by acoustic measurements. JASA 41, 1002-1010, 1967

Scripture, E. W.: The Elements of Experimental Phonetics, New York/ London 1902

Scripture, E. W.: Referate. Zeitschrift für Experimental-Phonetik I (3/4), 171-88, 1932

Techmer, F.: Naturwissenschaftliche Analyse und Synthese der hörbaren Sprache. Intern. Zeitschr. f. allgemeine Sprachwissenschaft I, 69-170, 1884

Tillmann, Hans G. Das Subjekt und seine individuelle Identität im phonetischen Kommunikationsprozess. IPK-FB Band 48, Buske, Hamburg 1973

-- (TmM (mit Phil Mansell)): Phonetik -- Lautsprachliche Zeichen, Sprachsignale und der lautsprachliche Kommunikationsprozess. Klett-Cotta, Stuttgart 1980

-- (EmP): Early modern phonetics, especially experimental and instrumental work. In R. E. Asher & J. M. Y. Simpson (Eds.), *The encyclopedia of language and linguistics* (Vol. 6, pp. 3040–3051). Oxford: Pergamon. 1994/2005[39]

-- (Kohler_FS): Kleine Phonetik und Große Phonetik. Phonetica 52, 144-159, 1995

-- Why the word should become the central unit of phonetic speech research. ZAS Papers in Linguistics 11, 1-20, Berlin 1998

-- (Heike_FS): Von der kleinen zur großen Phonetik: Wie es an der Universität Bonn in der Mitte des letzten Jahrhunderts zu einem Paradigmenwechsel in der phonetischen Sprachforschung kam. Hallesche Schriften zur Sprechwissenschaft und Phonetik, Band 45, Frankfurt 2013

Tillmann, H. G., G. Heike, H. Schnelle und G. Ungeheuer: DAWID I – Ein Beitrag zur Automatischen Spracherkennung, Beitrag A12, 5. Intern. Akustikkongress, Lüttich 1965

Tillmann, H.G. & B. Pompino-Marschall: Theoretical Principles Concerning Segmentation, Labelling and Levels of Categorical Annotation for Spoken Language Database Systems, Proc. of EUROSPEECH 1993, pp. 1691ff, Berlin 1993

Ungeheuer, E.: Wie die elektronische Musik ´erfunden´ wurde..., B. Schott´s Söhne, Mainz 1992

Ungeheuer, G. : Elemente einer akustischen Theorie der Vokalartikulation. Berlin 1962

Vennemann, Th. und Joachim Jacobs: Sprache und Grammatik. Darmstadt 1982

Zierdt A., Hoole P., Honda M., Kaburagi T., Tillmann H.G.: Extracting tongues from moving heads, in Proc. of the 5th Seminar on Speech Production: Models and Data, pp 313-316, Kloster Seeon 2000

---

[39] Also p. 401-416 in Koerner et. al (1995)

Hugo Pipping – a pioneer phonetician
at the University of Helsinki

Reijo Aulanko

University of Helsinki, Institute of Behavioural Sciences, Phonetics
reijo.aulanko@helsinki.fi

Abstract: Hugo Pipping was the first person to officially function as a phonetics teacher at the University of Helsinki. His most important contribution to Finnish phonetics was the introduction of scientific physiological research methods into the study of speech. Especially, he used the Phonautograph/Sprachzeichner developed by Victor Hensen in Kiel to manually analyse the spectral structure of vowel sounds, mostly the vowels of his native Finland-Swedish language but also those of Finnish. Pipping held the position of "docent" of phonetics in Helsinki until 1903, when he was invited to the university of Gothenburg to the position of docent of Nordic languages and phonetics. After four years in Sweden, he returned to Helsinki in 1907, but this time as a professor of Swedish language and literature (later professor of Nordic philology). Pipping retired in 1929, but remained academically active until his death in 1944.

1. Introduction

Knut **Hugo Pipping** (5 November 1864 – 31 July 1944) was appointed the first teacher ("docent") of phonetics in Helsinki in 1891 [1–3]. At that time, the University was still called the Imperial Alexander University of Finland, and in the 19th century, several new disciplines were created as a result of a generally more scientific orientation of the University [4].

Before reaching this position, Pipping had, in the course of his student years that included even studies of English, psychology and social anthropology in London [2], come into contact with Prof. Hällsten at the Institute of Physiology in Helsinki in 1886. In the late 19th century, ordinary physiological research instrumentation also made it possible to study some topics concerning speech production, and phonetic applications of physiological measurements were a common trend in European phonetics. With a recommendation from prof. Hällsten, Pipping traveled to study physiological phonetic methods in Kiel under Victor Hensen, whose Phonautograph/Sprachzeichner he used in 1888–1889. The result of these experiments was Pipping's doctoral dissertation *Om klangfärgen hos sjungna vokaler* [On the timbre of sung vowels] [5], which he defended in Helsinki on 22 January 1890.

In connection with Pipping's appointment in 1891, a phonetics laboratory was also founded in the Institute of Physiology [3], and during his years as a fulltime phonetics teacher (1891–1903), Pipping also took care of this phonetics lab. Within the University of Helsinki, the "Historical-Philological Section" (predecessor of the modern Faculty of Arts) also started to pay attention to the importance of phonetic studies, and, in 1903, gave special funding for the purchase of phonetic equipment (e.g. a kymograph). But at that time Pipping's career was already moving towards philological topics and positions, and the main responsibility for phonetic instrumentation fell on Pipping's successor as teacher of phonetics, the Frenchman Jean Poirot (1873–1924), who held that position from 1903 to 1920.

In the following, the phonetic studies of Hugo Pipping are described in some detail. To a modern reader, Pipping's articles seem to contain a surprisingly high amount of unprocessed raw data. For example, the publication *Über die Theorie der Vocale* (1894) contains only 30 pages of text, followed by a 30-page table of individual measurement values of the waveforms, plus a 4-page mathematical appendix written by Dr. Ernst Lindelöf.

2. Acoustic studies with Hensen's Sprachzeichner

Victor Hensen's Sprachzeichner or Phonautograph was one of the early methods of transforming soundwaves into visible form [6]. It consisted of a mouthpiece into which the subject spoke, a vibrating membrane at the far end of the tube, and a sharp diamond to register the movements of the membrane on to a glass plate. The curves on the glass plates were then read and measured with a microscope [5, 1]. All Pipping's early publications show the laboriousness of the time-consuming manual spectral analysis; it was already an achievement to manually measure the amplitudes and frequencies of the components of a single vowel sound, period by period [6]. But in spite of the thoroughly physical nature of the experiments he performed, Pipping saw (and especially liked to present) the Sprachzeichner primarily as a "tool for linguistics" [7].

2.1. Doctoral dissertation in 1890 (*Om klangfärgen hos sjungna vokaler*)

As the topic of his doctoral dissertation [5] Pipping chose acoustic measurements of the timbre of sung vowels. In this study, Pipping was more interested in the physical structure of vowels in general, rather than in the vowels of a specific language or dialect. The material consisted of a total of eight different vowels. Seven isolated vowels (as in the example words *lag, år, är, öl, yr, ed, is*) were sung by Pipping himself and by his wife (A. P.), both representing the Finland-Swedish dialect spoken by the educated classes in Helsinki. The only non-Swedish vowel studied was the vowel [u] ("as in German *du*") sung by "Miss C. S-P." and by "Dr. W. M.", both from Kiel. In early vowel studies like this, sung (rather than spoken) vowels were often used in order to obtain a less variable production of a given vowel quality, not because of special interest in singing.

A German translation of the dissertation – with an additional "Nachtrag" – was printed in *Zeitschrift für Biologie* [8, 9]. In both the Swedish and German versions example waveforms are presented of the vowels studied (see Fig. 1). Pipping's [8] description of the way in which the harmonic partials of the vowels are affected by resonances still sounds relatively adequate:

- In Uebereinstimmung mit Helmholtz habe ich gefunden, dass jeder Vocal sich durch ein oder mehrere Verstärkungsgebiete constanter Tonhöhe auszeichnet. Die Intensität seines Theiltones ist, caeteris paribus, grösser, je genauer er mit dem Maximalpunkte eines solchen Verstärkungsgebietes zusammentrifft.

Pipping [5] summarized his general observations in four statements:

1) sung vowels are composed exclusively of harmonic partial tones;

2) the intensity of a harmonic does not correlate significantly with its number in the order;

3) vowels differ from each other with regard to the number and width of the resonance areas and to their position on the frequency scale;

4) in many cases, different individuals speaking the same dialect produce nearly identical renderings of the same vowel.

On the basis of the data reported by Pipping [5, 8], it is not possible to represent the complete acoustic vowel system in a F1/F2 diagram. The following tendencies, however, were observed [10]:

[u]: the two "amplification areas" (förstärkningsområden) are at about 265 Hz and 800 Hz.
[i]: the characteristic frequencies are about 320 Hz and 2200 Hz.
[y]: the characteristic frequencies are about 340 Hz and 2150 Hz.
[ø]: the characteristic frequencies are about 400 Hz and 1060 Hz.

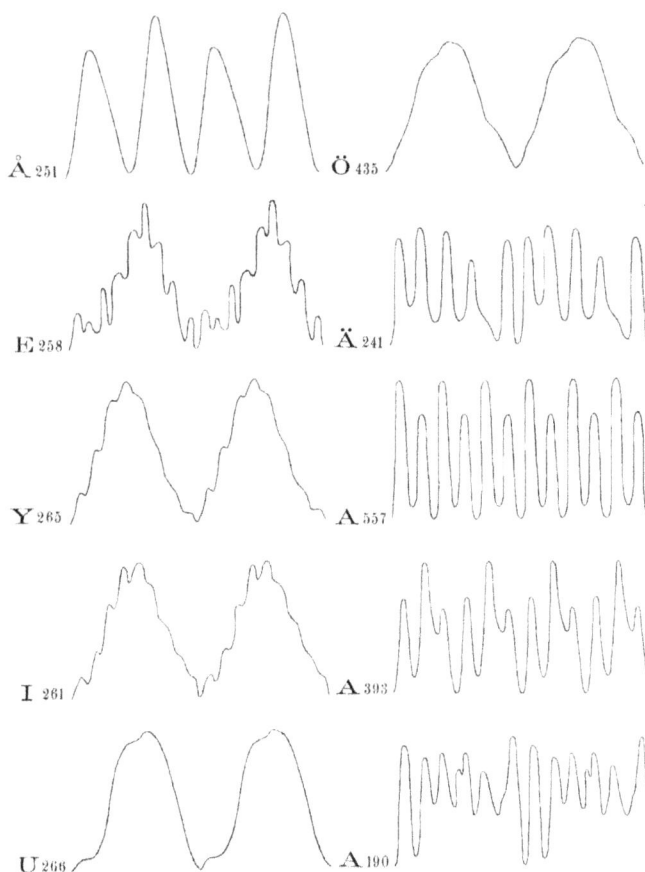

Figure 1: Pipping's [5] "original curves" of eight different vowels, the three A-curves on the right showing that "the resonance tone of a vowel remains the same even if the fundamental frequency varies". (The letter symbols have the following IPA values: Å [o], E [e], Y [y], I [i], U [u], Ö [ø], Ä [æ], A [ɑ].)

[e]:   the characteristic frequencies are about 350 Hz and 2350 Hz.

[ɑ]:   the characteristic frequency is about 1140 Hz, and a secondary amplification, of minor importance, is at about 2300 Hz. (In today's terms, the lower frequency would perhaps be F2, and the higher F3; i.e. Pipping did not find a "proper F1".)

[æ]:   the characteristic frequency is about 1400 Hz, a secondary peak about 2400 Hz. (Thus, also with this vowel, Pipping failed to find the correct F1.)

[o]:   the characteristic frequency is about 520 Hz. (No other resonance maximum could be found by Pipping.)

Later in the same year, Pipping published another article [7], where he was concerned with the applicability and usefulness of the phonautographic method to linguistics. In this study Pipping's topic was not the timbre of vowels but the quantities of speech sounds and the "musical and expiratory accents" in Finland-Swedish. More results on the same topic were published four years later [11], this time illustrated with pitch curves of the test words and sentences.

2.2. Second trip to Kiel

After spending one more year in Kiel (1893–1894), Pipping published two new studies on vowels. The material in these articles included the ten vowels of the Swedish dialect spoken in Helsinki. All the ten vowels were sung by Pipping and his wife, and five of them were also spoken by H. P. In the first of these studies [12], Pipping discussed the stable and variable characteristics of different vowels and presented the results of some new measurements. Pipping reflected upon the possible articulatory and/or acoustic differences in men, women and children producing a given vowel, and stressed the importance of determining how the human ear sorts out the relevant characteristics for vowel identification. The publication also contained a four-page mathematical appendix written by Ernst Lindelöf. Figure 2 reproduces Pipping's "vowel prism" which shows the different vowels in a diagram with the horizontal and vertical axes representing "Abstimmung des vorderen Resonanzraumes" (i.e. F2) and "Resonanzton des hinteren Raumes" (F1), respectively.

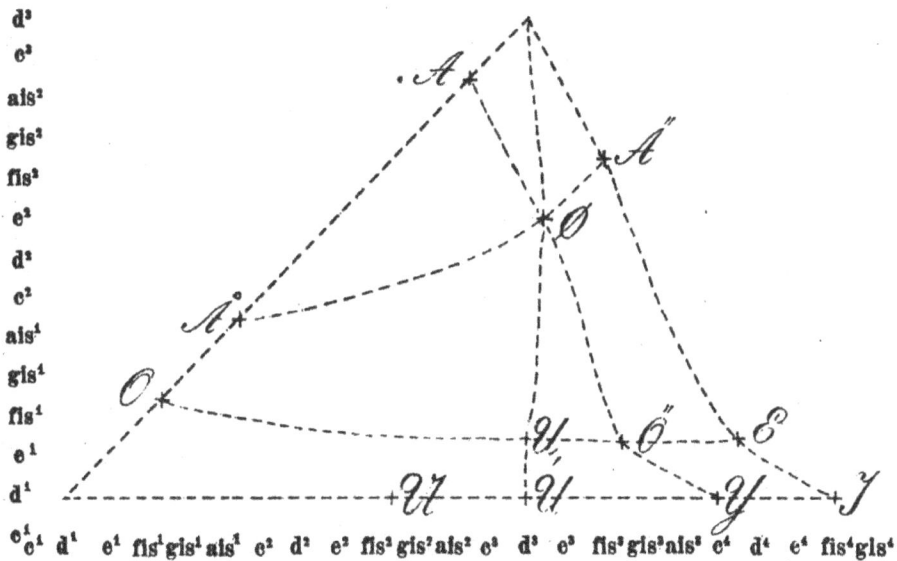

Figure 2: Pipping's [12] "vowel prism". The horizontal axis represents the resonance of the "front cavity", the vertical axis that of the "back cavity". The letter symbols have the following IPA values: Å [o], O [u], E [e], Y [y], I [i], Ö [ø], Ä [æ], A [ɑ]; "U" and "U₁" are two variants of the Finland-Swedish [ʉ], and the leftmost symbol on the bottom line is the German [u]. It should be noted that the vowels [u], [o], and [ɑ] fall on the diagonal, i.e. their F1 and F2 lie very close to each other.

In the second study [13] Pipping presented the results of analysing altogether 177 curves. He also dealt with the perception of fundamental frequency in vowel sounds. The measurements reported were actually only average values of the raw data published in [12]. In general, the material gives a much better chance to find the resonance maxima than the data analysed in earlier studies, because the fundamental frequency used in singing the vowels is now lower. Consequently, Pipping is, for instance, able to separate two peaks in the vowel [ɑ], where only a single wide resonance area was seen previously. Figure 3 shows the measured vowel resonances ("Resonanzhöhen und Resonanzbreiten") [13]. According to Pipping, the characteristic tones (≈ formants) tend to be slightly higher in spoken than in sung vowels.

Pipping also gave a very brief comparison of the Finland-Swedish vowels with the German ones, but in auditory terms only.

A					gis^3_6	cis^3_8						
E	f^1_{12}							fis^3_4		cis^4_1		
I	d^1_{12}									cis^4_3	fis^4_1	
O		g^1_{12}										
U	$d^1_{12}-f^1_{12}$						d^3_5					
Y	d^1_{12}								c^4_1			
Å		h^1_{18}										
Ä			g^2_5				fis^3_{10}					
Ö	f^1_{12}							g^3_5				
Ø			e^2_6				dis^3_7					

Figure 3: Heights and widths of vowel resonances from Pipping [13]; the horizontal axis is like a musical scale from low to high notes; "c^4_1", for instance, indicates that the resonance maximum is on the note c^4 (= 2112 Hz) and has a width of 1 semitone. (The letter symbols have the following IPA values: A [ɑ], E [e], I [i], O [u], U [ʉ], Y [y], Å [o], Ä [æ], Ö [ø], Ø [œ].)

2.3. *Zur Phonetik der Finnischen Sprache*

Nine years after his doctoral dissertation, and on the initiative of the Finno-Ugrian Society [1] Pipping published a study [14] on "the phonetics of Finnish", consisting mainly of acoustic analyses of spoken and sung Finnish vowels and some short notes on certain consonants and prosodic phenomena ("musical and dynamic accents"). In this work, Pipping's aim was to analyse and describe all Finnish vowels (short and long) and diphthongs spoken by one informant, and to shed light on the dialectal differences of Finnish by analysing the vowels sung by five male informants from different parts of the country. In addition to the differences in timbre, Pipping also intended to measure the durations of the vowels.

Recordings of the **sung** vowels were made in January–February 1896. According to Pipping [14], the number and position of resonances can be measured the more accurately, the more harmonic overtones there are, i.e., the lower the fundamental pitch is. In this respect, Pipping considered his optimal material to be the vowels sung on the note G# (Gis = 104 Hz) by "Mr. O. Nevalainen (bass) from Nurmes". However, although the resonance maxima reported for these vowels form a neat vowel system as such, they give the impression of an unnaturally reduced vowel space [10] (see Table 1).

Contrary to his wishes, Pipping could not draw any conclusions about dialectal differences on the basis of the sung vowels. What seemed to be decisive were the individual differences between the informants, e.g. between tenor and bass voices [14]. On the other hand, it is possible that, even when the singers have similar voice ranges, sung vowels need not necessarily reflect dialectal differences.

Table 1: The average values (in Hz) of the resonance maxima for the eight Finnish vowels sung by O. Nevalainen deduced from the data reported by Pipping [14].

Vowel	F1	F2
[ɑ]	492	1040
[æ]	492	1303
[o]	412	1009
[ø]	412	1310
[e]	412	1456
[u]	208	830
[y]	208	1432
[i]	208	1641

The results for the **spoken** vowels show a more interesting pattern. For the different vowels in spoken words, the speaker was "Mr. E. Ekman (tenor barytone) from Längelmäki". In the production of the following example words, Mr. Ekman's fundamental frequency varied from 98 Hz to 266 Hz [14]:

[ɑ]:   *satama, saadaan, taide, hauskuus*;
[æ]:   *pöytään, keihäitä, käytös*;
[o]:   *riemuitkoon, keino, neuvoin, Kuopio, houreet*;
[ø]:   *lyököön, käytös, pöytään, löit*;
[e]:   *siteet, houreet, keltä, kelta, tiede, taide, riemuitkoon, keihäitä, keino, neuvoin*;
[u]:   *hauskuus, kiuru, Kuopio, neuvoin, houreet, riemuitkoon*;
[y]:   *myllyyn, viipyi, käytös, pöytään, lyököön*;
[i]:   *viipyi, kiuru, neuvoin, Kuopio, keihäitä, löit, riemuitkoon, keino, taide*;

For the spoken vowels, Pipping tried to measure three different resonances for each vowel, the "thoracic", "pharyngeal", and "oral" resonances. The pharyngeal and oral resonances are here taken to correspond to the formants F1 and F2 in present-day terms. Unfortunately, Pipping does not present a clear summary of the resonance frequencies, but one interpretation of the results for the spoken vowels is shown in Table 2 and Figure 4. (The values of each vowel as the second component of a diphthong have been excluded from the averages calculated for theses presentations.)

Pipping [14] also made some observations on the effects of neighbouring consonants on vowel formants: A dental consonant makes the pharyngeal resonance (F1) descend and the oral resonance (F2) ascend, whereas a labial consonant makes the oral resonance descend.

Table 2: The average values (in Hz) of the "pharyngeal and oral resonances" (= F1 and F2) for the eight Finnish spoken vowels calculated from the data given by Pipping [14].

Vowel	F1	F2
[ɑ]	740	1220
[æ]	650	1560
[o]	510	940
[ø]	490	1600
[e]	470	1840
[u]	340	810
[y]	260	1780
[i]	310	1910

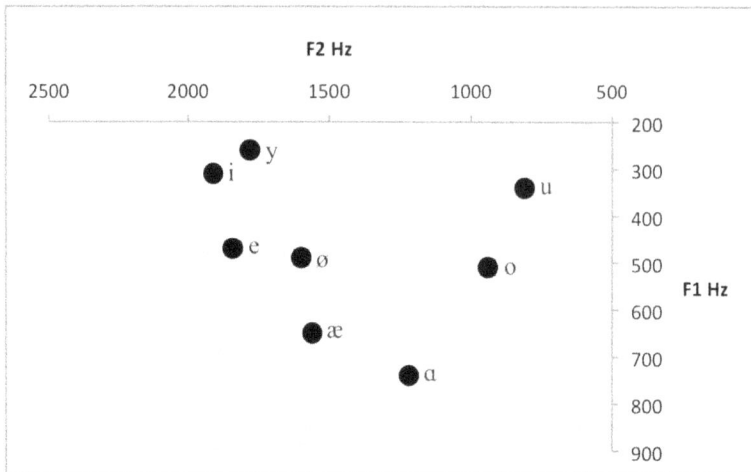

Figure 4: The Finnish vowels spoken by Mr. E. Ekman. Average formant values are shown in Table 2 [14, 10].

3. Other phonetic and philological studies

In spite of moving largely to philological studies after 1903, Pipping maintained his wide interests in phonetic topics, publishing articles on, e.g., the sound structure of Nordic languages, and especially on questions of metre. Pipping's later studies include a paper published in 1913 [15], where he dealt with the effects that the phase differences of overtones might have on the perception of timbre. Pipping wanted to draw conclusions on the functioning of the tympanic membrane of the human ear on the basis of the behaviour of the membrane of Hensen's Phonautograph. Among Pipping's later publications that lie somewhat further away from the phonetic focus, are his numerous philological articles on runic inscriptions and Edda poetry.

4. Pipping's impact on phonetics in Finland

It is generally held that experimental phonetics was introduced to Finland by Hugo Pipping [1–3, 16]. The impact that Pipping had on the phonetic (and philological) field in Finland can also be estimated from the impressive list of his publications that was included in his Festschrift (*Festskrift tillägnad Hugo Pipping på hans sextioårsdag den 5 november 1924*, Helsinki 1924), which was edited by his son Rolf Pipping, the list of publications compiled by his other son Hugo E. Pipping. As a phonetician, Pipping concentrated on strictly empirical and experimental methods, and even in his philological studies he maintained the phonetic point of view.

It is noteworthy that all through his career Pipping tried to bridge the gap between the methods of natural sciences and the humanistic problems of linguistics and philology [1]. He even envisaged the possibility of a physical understanding of historical sound changes [7, 17] on the basis of "acoustic analysis, which will preserve its value through millennia" (*...den akustiska analysen bibehåller sitt värde genom årtusenden och sålunda i framtiden kommer att möjliggöra en återblick öfver språkets utveckling*).

References

[1]   Enkvist, N. E.: Hugo Pipping som fonetiker. *Finsk Tidskrift* 156 (1954), 6–17.

[2]   Panelius, O.: *Hugo Pipping*. Helsingfors: Svenska Litteratursällskapet i Finland (1955).

[3]   Aalto, P.: *Modern Language Studies in Finland 1828–1918*. Helsinki: Societas Scientiarum Fennica (1987).

[4]   Klinge, M. & al.: *Keisarillinen Aleksanterin Yliopisto 1808–1917*. (Helsingin Yliopisto 1640–1990 II) Helsinki: Otava (1989).

[5]   Pipping, H.: *Om klangfärgen hos sjungna vokaler. Undersökning utförd vid fysiologiska institutet i Kiel medels Hensens fonautograf*. Helsingfors: J. C. Frenckell & Son (1890).

[6]   Schneider, A.: Change and continuity in sound analysis: A review of concepts in regard to musical acoustics, music perception, and transcription. In: R. Bader (ed.) *Sound—Perception—Performance*. Cham: Springer (2013), 71–111.

[7]   Pipping, H.: *Om Hensens fonautograf som ett hjälpmedel för språkvetenskapen*. Helsingfors: J. C. Frenckell & Son (1890).

[8]   Pipping, H.: Zur Klangfarbe der gesungenen Vocale. *Zeitschrift für Biologie* 27 (1890), 1–80.

[9]   Pipping, H.: Nachtrag zur Klangfarbe der gesungenen Vocale. *Zeitschrift für Biologie* 27 (1890), 433–438.

[10]   Aulanko, R.: The first 100 years of acoustic vowel studies at the University of Helsinki. In: R. Aulanko & M. Leiwo (eds.) *Studies in Logopedics and Phonetics 2*. Publications of the Department of Phonetics, University of Helsinki, Series B 3 (1991), 9–32.

[11]   Pipping, H.: Fonautografiska studier. *Finländska bidrag till svensk språk- och folklifsforskning utgifna af Svenska Landsmålsföreningen i Helsingfors*, 99–110.

[12]   Pipping, H.: *Über die Theorie der Vocale*. (Acta Societatis Scientiarum Fennicæ XX:11) Helsingfors: Societas Scientiarum Fennica (1894).

[13]   Pipping, H.: Zur Lehre von den Vocalklängen. *Zeitschrift für Biologie* 31 (1895), 524–583.

[14]   Pipping, H.: *Zur Phonetik der finnischen Sprache. Untersuchungen mit Hensen's Sprachzeichner*. (Mémoires de la Société Finno-Ougrienne 14) Helsingfors: Société Finno-Ougrienne (1899).

[15]   Pipping, H.: *Studien über die Funktion des Trommelfells*. (Acta Societatis Scientiarum Fennicæ XX:11) Helsingfors: Societas Scientiarum Fennica (1894).

[16]   Iivonen, A.: An outline of the history of phonetics and logopedics at the University of Helsinki. In: M. Leiwo & R. Aulanko (eds.) *Studies in Logopedics and Phonetics 1*. Publications of the Department of Phonetics, University of Helsinki, Series B 2 (1990), 7–15.

[17]   Rischel, J.: The contribution of the Nordic countries to historical-comparative linguistics: Rasmus Rask and his followers. In: O. Bandle & al. (eds.) *The Nordic Languages. Volume 1*. Berlin & New York: Walter de Gruyter (2002), 124–133.

A brief history of articulatory-acoustic vowel representation

Coriandre Vilain*, Frédéric Berthommier, Louis-Jean Boë

- Univ. Grenoble Alpes, GIPSA-Lab, F-38000 Grenoble, France
- CNRS, GIPSA-Lab, F-38000 Grenoble, France

(*) corresponding author: coriandre.vilain@gipsa-lab.fr

Abstract: This paper aims at following the concept of vowel space across history. It shows that even with very poor experimental means, researchers from the 17th century started to organize the vowel systems along perceptual dimensions, either articulatory, by means of proprioceptive introspection, or auditory. With the development of experimental devices, and the increasing knowledge in acoustic and articulatory theories in the 19th century, it is shown how the relationship between the two dimensions tended to tighten. At the mid 20th century, the link between articulatory parameters such as jaw opening, position of the constriction of the tongue, or lip rounding, and the acoustical values of formants was clear. At this period, with the increasing amount of phonological descriptions of the languages of the world, and the power of the computer database analysis allowing extracting universal tendencies, the question of how the vowel systems are organized arose. The paper discusses this important question, focusing on two points: (1) how the auditory constraints shape the positioning of a specific set of vowel within the acoustic space, and (2) how the articulatory constraints shape the maximal extension of the vowel systems, the so-called maximal vowel space (MVS).

1. Introduction: articulatory or auditory representation of vowels?

In the past centuries, the vowel space representation evolved along two main dimensions: articulatory and auditory. One of the first articulatory representations of vowels was proposed by Robinson in 1617. By capturing the position of his tongue positions during vowel production, he proposed to categorize the vowels along the anterior-posterior position of the tongue (see [1] for details). More than a hundred years later, Hellwag proposed in turn, one of the first triangle representations of the German vowel space with its degrees and its order relations ([2: §57], see fig. 1 left). Now, the auditory perception seemed to be of importance in this description. As quoted by Nearey: « Hellwag is partly interested in anatomical descriptions of speech production. However, his introduction of the triangle seems to be motivated primarily by concern for some type of auditory, rather than articulatory relationship» ([3: 41]). Indeed, Hellwag was much influenced by Reyher's 1679 works on "tone heights", defined from whispered vowels and that relied on the timbre of the vowels independently of any fundamental frequency (see [1]).

At the same period, the German physicist Chladni ([4]) proposed another representation of the 10 vowels, that he claimed to be the whole set of vowels, sorted into 3 series according to articulatory considerations[1] ([4: § 52], see fig. 1 right). From the vowel /a/, he derived a 1st

---

[1] It is interesting to note that this description of vowels is absent from the original 1802 German edition, but it is present in the 1809 French expanded edition.

branch with "open outside and slowly narrowing inside" (/a ɔ o u/), a 2nd branch with an identical "open outside and slowly narrowing inside"[2] (/a ɛ e i/), and the 3rd branch with "slowly narrowing outside and inside" (/a œ ø y/). However, at this time, these articulatory descriptions remained quite qualitative.

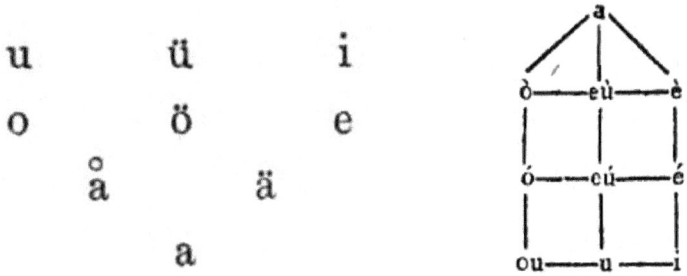

Figure 1. First vowel triangle representations. On the left, Hellwag ([2: § 57); on the right, Chladni ([4: §52]).

A formalization of the vowels classification in terms of more quantitative horizontal and vertical tongue position has been proposed in 1867 by A.M. Bell in his famous *Visible Speech* ([5]). In Bell's system, specific graphical symbols coding for the positions of the speech articulators were defined. These symbols were used as supports to help deaf people in learning to speak. H. Sweet further developed Bell system to give rise to the Bell & Sweet model ([6]). In this model, the explicit link between articulatory configurations and vowel classification was made, using introspective sensations of the tongue position instead of auditory assessment. Indeed, as quoted by Catford, Sweet gave more credit to articulatory than auditory skills to classify vowels: "there can be no question that flexible organs well trained together with only an average ear, can yield better results than even an exceptionally good ear without organic training". ([7: 22]).

Following Bell & Sweet, Passy, founder of the IPA in 1886, explicitly arranged the natural language vowels along two articulatory dimensions: (1) front-back, with three degrees of articulation (palatal, mixed and velar) and (2) open-close, with four degrees of opening ([8],[9], see fig. 2).

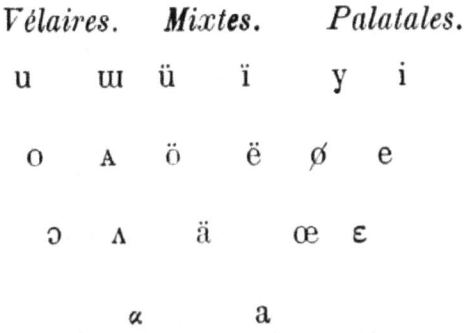

Figure 2. Natural vowels representation by Passy in 1888 (extract from [8])

---

[2] No reference was given to the lip position in the first 2 branches, which would allow distinguishing between them.

It is noteworthy that in Passy's description, the arrangement along the front-back dimension seems to be associated with the acoustical dimension of « clarity », as attested by the vicinity of /i y/, /i u/, and /ɯ u/ [3: 53]).

In the continuation of Passy's work, Jones introduced in 1917 his primary cardinal vowels containing 4 front unrounded vowels /i/, /e/, /ɛ/, /a/ and 4 back vowels, the unrounded /ɑ/, and the 3 rounded /u/, /o/, /ɔ/ ([10], [11]). These cardinal vowels were defined as a set of references to be compared with the vowels of the actual languages of the world. They were supposed to limit the extension of these vocalic systems. In Jones's view, the 1st (/i/) and the 5th (/ɑ/) vowels were the only ones that could be articulatory defined; the others were just auditorily defined in perceptually equidistant steps. The secondary cardinal vowels were defined with an opposite rounding feature: the 4 front rounded vowels /y/, /Ø/, /œ/ /œ/, the 3 back unrounded vowels /ɯ/, /ɣ/, /ʌ/, and the back rounded vowels /ɒ/. It is worth noting that in Jones, and more generally in the IPA view, taking into account the lips rounding explicitly allowed discriminating vowels with the same place of articulation and the same aperture (fig. 2).

To summarize, up to the early 20th century, the articulatory representation tended to be the primary means to organize and classify the vowels in a convenient way. Even if the acoustical dimension was not completely absent, it remained of poor importance since the human ear was the sole means of analysis of the speech sounds. This changed at the end of the 19th century with the development of acoustic theories, and of devices dedicated to record, measure and analyse these speech sounds.

2. The acoustic measurements of formants

The notion of formants is intrinsically connected to the notion of acoustic resonance in tubes. A large amount of research in acoustics had been performed in the early 19th century. Chladni, for instance, is an icon in this field of research, with his pioneer work on vibrations of plates and the famous Chladni figures they are associated with. But interestingly, not much was carried out to understand sound propagation in the vocal tract. Willis was one of the first to consider this question ([12]).

2.1. Emergence of the notion of formant

In 1830, Willis proposed that the vocal tract could be modelled as an acoustical tube with a natural frequency, directly given by its length. It led him to arrange the vowels as a function of their corresponding tube length in a /i e a o u/ sequence ([12], cited in [13: 46]). Then, Wheatstone, in 1837, detailed how the acoustic resonator modifies the spectral properties of the sound emitted by a vibrator: « when an air resonator is approached by a vibrator, the sound of the latter is considerably reinforced if the natural frequency of the resonator coincides with the frequency of one of the harmonics of the vibrator » (13: 47). Helmholtz, in 1863, systematized the idea that harmonics of the glottis source are reinforced by the vocal tract resonances and he assigned to each vowel one or two particular frequencies that characterize them: the *vocables* ([14]). These *vocables* corresponded, according to him, to the natural frequencies of the mouth cavities. In his approach, the back vowels /u/, /o/, /a/ were characterized by one *vocable* only (indeed, for these vowels the first two formants are too close to be separated by Helmholtz's measurement apparatuses). The front vowels /i/, /œ/, /e/, /æ/, at the contrary, were characterized by two *vocables*.

Interestingly, Hermann was the first to propose the term of *formant* to describe the resonance frequencies of the vocal tract in 1894 ([15]). But he failed to accurately compute formant values, due to his misunderstanding of the relationship between the harmonics produced by the glottal source and the resonance frequencies of the vocal tract. This issue has been solved lately when the acoustic measurements were able of discriminating all the speech signal frequencies and not only the reinforced harmonics as in Hermann's time. For instance, the development of electrical synthesis devices capable of reproducing vowel sounds has been of great help for better understanding their structural properties. In this domain, the first full electrical synthesis device was proposed by Stewart in 1922 ([16]). His synthesizer had a buzzer for the excitation and two resonant circuits to model the acoustic resonances of the vocal tract (see fig. 3). The apparatus was only able to generate single static vowel sounds with their two lowest formants. With a similar, albeit upgraded four resonant circuits apparatus, Obata & Teshima discovered the third formant in vowels in 1932 ([17]).

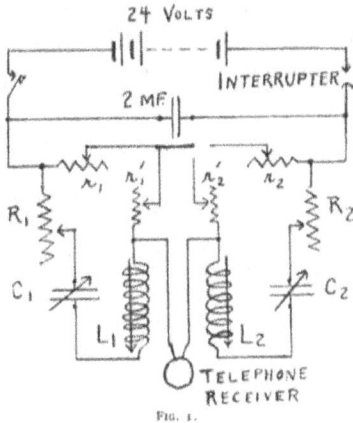

Figure 3. Stewart's electrical analogue of the vocal tract with two resonant circuits corresponding to the two formants (from [16]).

With the definition of the first three formants, the vowels could be acoustically classified in a convenient way, due to the matching between the perceptive distance in the auditory domain and the spectral distance in the F1-F2-F3 space. The main concern, in the late 19th and early 20th century was to develop experimental devices, capable of precisely measuring the formant values.

2.2. Emergence of the links between formant and articulatory measurements

Helmholtz was the first to propose reliable measurements of the reinforced harmonics in an acoustical tract ([14]). In France, Kœnig developed his *"Analyseur à flammes manometriques"* (manometric flames analyser) based on a bench of Helmholtz resonators connected with manometric valves, allowing for the visualization of the air vibrations by use of visible "dancing flames" (see fig. 4). By this means, he was able to grossly describe the spectral content musical instrument sounds, as well as sung French vowels ([18: 61], cited in [19]).

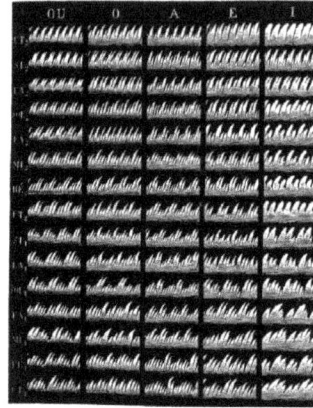

Figure 4. Kœnig's manometric flames analyser (left) and graphical inscription of the French vowels (/u/, /o/, /a/, /e/, /i/) "dancing flames" (right).

In 1926, Stumpf found the vowel resonances given by Helmholtz with oscillographic measurements ([20]). Independently, Crandall & Sacia ([21]) followed by Paget ([22]) used a photo-mechanic device to compute the vowel spectra and succeeded in measuring the first two formants of each vowel (except for /a/). Essner ([23]) analytically computed the vocal tract resonances and presented once again (more than a century after Hellwag), the vowels in the F1-F2 space. However, he did not explicitly link the acoustical and the articulatory representations. Joos ([24]) and Delattre ([25]) are the ones who bound together, for the first time in 1948, these two representations into a common framework as they associated the IPA articulatory quadrangle with the acoustic F1-F2 space. As Joos remarked: "the correlation between articulation and vowel color is [...] astonishly simple. Although the vowel samples have here been placed on the chart strictly according to acoustical measurements (made from a phonograph record!) the diagram is practically identical with the 'tongue position chart" (24: 53). In Delattre's paper, the horizontal axis corresponds to F2 value and it is explicitly associated with the front-back position of the tongue and with the labialisation. The vertical axis corresponds to the F1 value, associated with the aperture of the vocal tract (See fig. 5). For these two cases, it has to be noted that the emergence of spectrographic measurements, allowing for direct measurements of formant was of crucial importance.

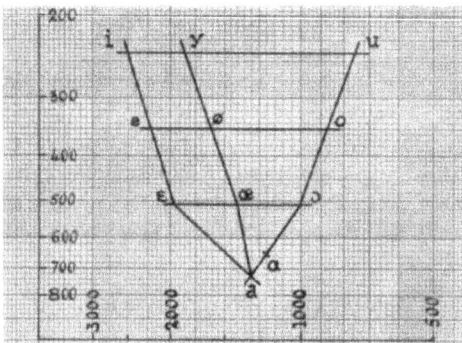

Figure 5. Vocalic triangle of the French vowels with F1 and F2 values measured from spectrograms (from [25]).

3. The modern era: towards a better comprehension of the relationship between articulatory and acoustical representations

3.1. The development of articulatory synthesis

As previously mentioned, Stewart proposed in 1922 the very first electrical device allowing synthesizing vowels ([16]). In his apparatus, the electrical analogue of the vocal source was a simple buzzer and the vocal tract is modelled by two resonant circuits in parallel, each of them being tuned independently to correspond to the first two formants (see fig. 3). The main issue pointed by the author was not the construction of the device itself but « the manipulation of the apparatus to imitate the manifold variations in tone which are so important in securing naturalness » ([16]). Indeed, there was no real articulatory model in this very first apparatus, just an ad hoc control of the electrical characteristics allowing to produce « realistic » sounds to the most possible extent, while exploring the acoustic space.

Chiba & Kajiyama ([13: §10]) elaborated the first articulatory model from the area function of the vocal tract. Moreover, they synthesized vowel sounds using replicas of the vocal tract and a larynx-tone emitter, and they compared the spectra of synthetic and natural vowels. Given this articulatory model, they considered the problem of *affiliation*, i.e. the identification of the articulatory configuration yielding a given formant value for vowels.

In this period, the development of transmission line theory that model the vocal tract as a succession of elementary electrical cells, that match the acoustical properties of the elementary acoustical tube, made rapid progress and "provided an opportunity to reproduce and to control the many articulatory positions and movements that occur in speech" ([26: 741]).

Dunn ([27]) proposed in 1950, a 4-tubes analogue of the vocal tract that allowed him computing analytically the first three acoustical formants (fig. 6). He demonstrated then his electrical vocal tract (EVT) composed of 25 cells, each representing a cylinder 0.5 cm long and 6 cm3 in cross section plus a variable inductance that can be inserted at any position between 2 sections of the line, representing the "tongue hump constriction". Only, the tongue hump characteristics could be controlled in this EVT, by means of 3 parameters: Constriction Place ("position of the tongue hump"), Constriction Area ("magnitude of the tongue hump") and Lip Area ("magnitude of the lip constriction"). Indeed, the limitations of such a device were important as pointed by the author himself: "the whole series of English vowels can be produced by this apparatus – not perfectly, but distinctly better than we were able to make with three independent tuned circuits without additional suppression between and above the resonances" ([27: 751]).

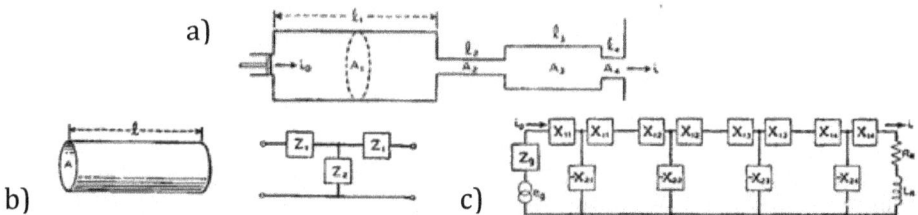

Figure 6. Dunn's theoretical model of vocal tract (from [27]). a) The 4 tubes approximation. b) The elementary cell corresponding to a cylinder of section A and length L. c) The complete model including the 4 elementary cells, an electrical analogue of the glottal source on the left and an electrical analogue of the acoustic radiation impedance on the right.

Stevens et al. added: " Only a limited number of positions of the vocal tract can be simulated accurately by this small number of controls" and "the approximation of two cylinders of uniform cross-sectional area with lumped tongue and lip constriction is likely to be an error in considerable margin" ([26:734-735]).

Stevens et al. ([26]), following Dunn's work, realized their own electrical analogue of the vocal tract composed of 35 sections of 0.5 cm each with different cross-sections corresponding to a discretization of the vocal tract areas based on x-ray pictures published by Polland & Hala in 1926 ([28]) and Russel in 1928 ([29]). With this device, "it has been possible to synthesize all the English vowels. The quality of the vowels is judged by experienced listeners to be good" ([26: 739]). However, all the elementary cells had to be controlled independently. Stevens & House aimed at combining the simplicity of Dunn's approach and the accuracy of Stevens et al 's device ([30]). They established "relations between the parameters that describe the articulatory events and the formant frequencies that describe the acoustic events" ([30: 484)). In this way, they defined the nomograms that associate resonance modes (here, the first three formant frequencies) to a given vocal tract configuration (here, the following three articulatory parameters: the degree of constriction (r0), the place of constriction (d0), and the mouth opening (A/l)). By this means, they were able to compute the relationship between acoustical dispersion areas as measured by Peterson & Barney in 1952 ([31]).

Fant, in 1960 ([32]) took back the 4-tube model theoretically described by Dunn ([27]) and proposed a simplified model of the vocal tract: 2 cavities, one front and one back, of fixed dimensions separated by a constant length constriction, a lip area and a fixed lip length. The model was driven by 3 parameters: constriction position X_C, constriction area (A_C), and lip area A_L (see fig. 7). The nomograms allowed the measurements of F1 to F5 as a function of X_C, for a given set of A_C and A_L. With Fant's approach, one was able to explain the formant characteristics of all the oral vowels.

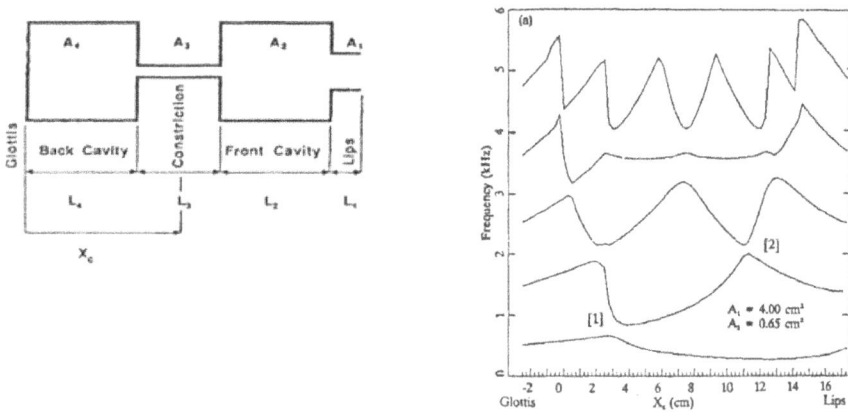

Figure 7. Left, the four-tube model proposed by Fant ([32]). Back Cavity (pharynx) : A4 = 8cm2, L4 can vary between 0 and 15cm ; Tongue constriction : A3 = 0.65 cm2, L3 = 5cm ; Front cavity (mouth) : A2 = 8cm3, L2 can vary between 0 and 15 cm ; Lips : A1 = 0.16 or 4 cm2 , L1 = 1cm. The constriction center coordinate Xc can vary from -2.5 to 17.5 cm, keeping L2+L3+L4 = 15cm. Right, a corresponding nomogram for a middle lip opening A1 = 4.0 cm2, A2 = 0.65 cm2.

154

Using Fant's nomograms, Gunnilstam ([33]) in 1974, followed by Badin et al. in 1988 ([34]) detailed a method to affiliate formants to the vocal tract cavities. Finally, with the matching between articulatory and acoustic parameters, the phonetic description of the languages of the world was then greatly facilitated.

3.2. The vowel system categorization and the maximal vowel space

Beyond an apparent diversity, the phonological typologies of the vowel systems have been shown for a long time to exhibit strong regularities (see [35]): /i a u/ for 3-vowel systems, /i e o a u/ for 5-vowel systems or /i e ɛ o ɔ a u/ for 7-vowel system for instance. In studies on large corpora, such as the UPSID database ([36]), allowed extracting universal tendencies in the vowel systems (see [37] for 209 languages, or [38] for the 417 languages described in UPSID). As stated by Liljencrants & Lindblom, it appeared there that "only a subset of all logically possible combinations of those formant frequencies is associated with formant vowels" ([41: 839]). The question of how the vowel systems are defined, on which bases, articulatory or auditory, was thus a question of importance in the late 20th century.

To answer this question, the principle of distinctiveness has long been claimed to be the correct criterion for vowel systems organization as it makes them easy to produce and to perceive. The formalisation of this principle in the perceptual domain has been carried out in the dispersion theory (DT), proposed by Liljencrants & Lindblom ([39]). The DT was the first quantitative simulation of vowel inventories. It was based on the maximization of the perceptual distances in the (M1, M2, M3) space, where Mi is the ith formant expressed in the Mel scale, to account for perceptual representations. The boundaries of the frequency domain in which the dispersion can occur, the maximal vowel space (MVS) is set empirically by computing from the phonological databases. By minimizing the criterion $\sum_{i=1}^{N} 1/r^2$, where r is the distance between the ith pair of vowels and N is the number of pairs in the considered system, the DT proposes an organization of the vocalic systems of variable size ranging from 3 to 12. Schwartz et al. ([40]) further developed the seminal proposal of Liljencrants & Lindblom ([39]) and propose a second criterion to define the dispersion of the vocalic systems. Their dispersion-focalization theory (DFT) attempted to predict vowel systems based on the minimization of an energy function summing two perceptual components: global dispersion, which is based on inter-vowel distances (as in the DT); and local focalization, which is based on intra-vowel spectral salience related to the proximity of formants. It has to be noted that none of these theory account for any "articulatory cost" that could also determine the vowel dispersions. They were, indeed, purely auditory.

To go deeper into the vocalic system modelling, the question of the MVS was of importance too. Liljencrants & Lindblom ([39]) chose an empirical definition of the MVS from typological studies, but other researchers tried to compute it with acoustical simulations of vocal tract systems.

A first attempt, in this latter direction, was proposed by Bonder ([41], [42]) who described an analytical method to compute the MVS for a 4-tubes-of-equal-length model. With this model, he proposed to solve the inverse problem, i.e. to get the vocal tract shape from the formant values, by using the minimal articulatory difference (MAD) method ([43]). He computed the relationship between the (F1, F2) values and the geometrical shapes of the vocal tracts within the MVS (see fig. 8).

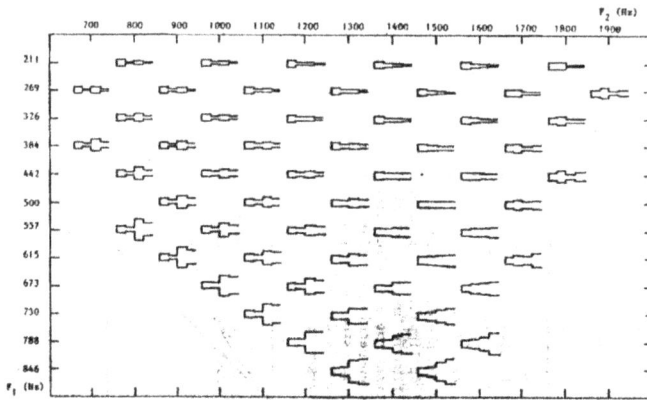

Figure 8. The solution of the inverse problem by means of the MAD model. Each tube has an overall length of 17.5 cm. The glottis is situated at the left side of the tubes (from [43]).

Finally, following Bonder and benefiting from the exponential growth of computer capacities, Boë et al. ([44]) proposed a stochastic approach to characterize the MVS with a n-tube model, with each tube having random dimensions, provided a constant global length. This geometrical/acoustic modelling of all the possible acoustical outputs as a function of the number of tubes allowed showing that, as long as n>3, all the MVS have a triangular shape in the F1-F2 plane. However, in the F2-F3 plane, the modelled MVS was much bigger than the one observed with real articulatory models of the vocal tract. So, whereas it is not the case in the F1-F2 plane, the articulatory constraints are of importance for defining F3. To define a somewhat realistic acoustic representation in the F1-F2-F3 plane, a numerical simulation based on an anthropological model is needed. Using the Maeda model ([45]), Boë et al. ([44]) succeeded in obtaining a realistic "horse-shoe" shaped MVS with accurate formant relationship for its boundaries (see fig. 9).

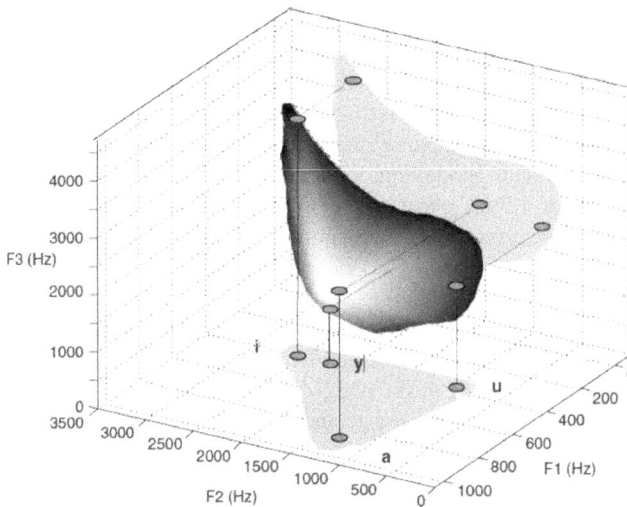

Figure 9. The F1–F2–F3 Maximal Vowel Space defined with an anthropological model. The four corner-point vowels in this space are [i y a u] (from [46]).

156

4. Conclusion

In this paper, we looked at the evolution of the vowels representations across history. If nowadays, the vocalic space is of common use, and easily taught as the common representation of articulatory and acoustic properties of vowels, its development had been far from trivial. From a historical standpoint, starting at the 17th century, we followed the building of the vocalic space, based on both the auditory and the articulatory dimensions. The major milestones in this evolution have been recalled, from the pioneers with their sole auditory and articulatory "impressions" as methods of investigation, to the very recent works benefiting from devices able to precisely measure acoustical and articulatory configurations of the vocal tract during vowel productions. We discussed how the acoustic measurements slowly emerged across history and how both acoustic and articulatory representations were unified in the mid 20th century. Given this unified view, some important questions on vowel systems organization arose. On this very topic, we focused on the two following: (1) how the auditory constraints shape the positioning of a specific set of vowel within the acoustic space, and (2) how the articulatory constraints shape the maximal vowel space. From the pioneer article of Liljencrants & Lindblom ([39]) to recent work including one of us ([44]), we showed that this question is still a matter of debate, and a major field of research in the current speech studies.

References

[1]   Pfitzinger H. & Niebuhr O. (2011) Historical development of phonetic vowel systems - The last 400 years. Proceedings of the 17th ICPhS, Hong Kong, China,160-163.

[2]   Hellwag C.F. (1781). Dissertatio de Formatione Loquelae. Heilbronn 1886 (French translation by Monin M.-P. available in Bulletin de la Communication Parlée, 1991).

[3]   Nearey T. M. (1977). Phonetic Feature Systems for Vowels. Indiana University Linguistics Club, 1978

[4]   Chladni E. F. F. (1809). Traité d'acoustique. Courcier, Paris.

[5]   Bell A.M. (1867). Visible Speech: The Science of Universal Alphabetics. Londres, Simkin, Marshall & CO.

[6]   Sweet H. (1877). A Handbook of Phonetics. Clarendon Press, Oxford.

[7]   Catford J.C. (1981). Observations of the recent history of vowel classification. In Asher, Henderson (eds.). Towards a history of phonetics. Edinburgh: The University Press, 19-32.

[8]   Passy P. (1888). Our revised alphabet. The Phonetic Teacher, 7–8: 57–60

[9]   Passy P. (1890). Étude sur les changements phonétiques et leurs caractères généraux, Paris, Librairie Firmin-Didot, 1890

[10]   Jones D. (1917). An English Pronouncing Dictionary. London: Dent.

[11]   Jones D. (1917). Speech Sounds — Cardinal Vowels (Short and Long). England, Hayes, Middlesex: The   Gramophone.

[12]   Willis, R. (1830). On vowel sounds, and on reed organ pipes. Transactions of the Cambridge philosophical Society, III, 231-276.

[13]   Chiba T. & Kajiyama M. (1941). The vowel, its nature and structure. Tokyo-Kaiseikan Publishing Company. Ltd, Tokyo.

[14] Helmholtz H. (1863). Die Lehre von den Tonempfindungen als physiologische Grundlage für die Theorie der Musik. Braunschweig, Vieweg.

[15] Obata J. & Teshima T. (1932). On the properties of Japanese vowels, Jap. J. Physics, Vol. 8.

[16] Hermann L. (1894). Beiträge zur Lehre von der Klangwahrnehmung. Pflügers Arch. 56, 467-499

[17] Stewart J. Q. (1922). An electrical analog of the vocal organs. Nature, 110, 311-312.

[18] Kœnig R. (1882). Quelques expériences d'acoustique, Imprimerie Lahure. Paris.

[19] Vilain C., Arnal A., Boë L.-J. (2011). L'analyseur de Koenig : un premier spectromètre pour l'étude de la parole, In Boë LJ & Vilain C. (Ed.) Un siècle de phonétique expérimentale : fondation et éléments de développement, ENS Editions.

[20] Stumpf C. (1926). Die Sprachlaute: Experimentell-phonetische Untersuchungen nebst einem Anhang über Instrumental-Klänge, Springer, Berlin.

[21] Crandall I.B. & Sacia C.F. (1924). A dynamical study of the vowel sounds. Bell system technical journal, 3(2).

[22] Paget R. (1930). Human Speech: Some Observations, Experiments, and Conclusions as to the Nature, Origin, Purpose and Possible Improvement of Human Speech, International Library of Psychology, Routledge and Kegan Paul.

[23] Essner C. (1947). Recherche sur la structure des voyelles orales. Archives néerlandaises de phonétique expérimentale, 20, 40-77.

[24] Joos M. (1948). Acoustic phonetics. Supplement to Language, 24(2), the linguistic society of America, Baltimore.

[25] Delattre P. (1948). Un Triangle Acoustique des Voyelles Orales du Français. The French Review, 21(6), 477-484.

[26] Stevens K.N., Kasowski S., Fant C. G. M. (1953). An electrical analog of the vocal tract. Journal of the Acoustical Society of America, 25 (4): 734–42.

[27] Dunn H. K. (1950). The calculation of vowel resonances, and an electrical vocal tract. Journal of the Acoustical Society of America, 22, 740-753.

[28] Polland P. & Hala B. (1926). Les radiographies de l'articulation des sons tchèques. (Artikulace českých zvuků v roentgenových obrazech), Praha.

[29] Russell G. O. (1928). The vowel, its psychological mechanism, as shown by x-ray. Columbus, OH: Ohio State University Press.

[30] Stevens K. N., & House A. S. (1955). Development of a quantitative description of vowel articulation. Journal of the Acoustical Society of America, 27, 401-493.

[31] Peterson G.E. & Barney H.L. (1952). Control Methods Used in a Study of the Vowels. Journal of the Acoustical Society of America, 24, 175–184.

[32] Fant G. (1960). Acoustic theory of speech production. Mouton, The Hague.

[33] Gunnilstam O. (1974). The theory of local linearity. Journal of Phonetics 2, 91–108.

[34]   Badin P., Boë L.J., Perrier P., Abry C. (1988). Vocalic nomograms: Acoustic considerations upon formant convergence. Bulletin de la Communication Parlée, 2, 65-94.

[35]   Troubetzkoy N. S., (1939), « Grundzüge des Phonologie », Travaux du Cercle Linguistique de Prague, Vol. 7, traduit en français par Cantineau J. en 1970, Principes de phonologie. Klincksieck, Paris.

[36]   Maddieson, I. 1984. Patterns of Sounds. Cambridge University Press, Cambridge. Paperback reprint 2009.

[37]   Crothers J. (1978). Typology and universals in vowel systems. In Universals of human language, (J. H. Greenberg, C. A. Ferguson & E. A. Moravcsik, editors) pp. 93–152. Stanford: Stanford University Press.

[38]   Schwartz J.-L., Boe L.-J., Vallée N., and Abry C. (1997). Major trends in vowel system inventories. Journal of Phonetics 25, 233–253

[39]   Liljencrants J. & Lindblom B. (1972). Numerical simulations of vowel quality systems: The role of perceptual contrast. Language, 48:839–862.

[40]   Schwartz J.-L., Boe L.-J., Vallée N., and Abry C. (1997). The dispersion-focalization theory of vowel systems. Journal of Phonetics, 25:255–286.

[41]   Bonder, L. J. (1983). The n-tube formula and some of its consequences. Acustica, vol. 52, no. 4, March, pp. 216–226.

[42]   Bonder, L. J. (1983). Equivalency of lossless n-tubes. Acustica, vol. 53, no. 4, August, pp. 193–200.

[43]   Bonder L.J. (1987). From formant space to articulation space by means of the mad model (personal communication)

[44]   Boë L.J., Badin P., Ménard L., Captier G., Davis B., MacNeilage P., Sawallis T.R., Schwartz J.L. (2013). Anatomy and control of the developing human vocal tract: A response to Lieberman, Journal of Phonetics, 41 (5), September 2013, 379-392.

[45]   Maeda S. (1979). An articulatory model of the tongue based on a statistical analysis.   J. Acoust. Soc. Am. 65, S22.

[46]   Schwartz J.L., Boë L.J., and Abry C. (2007). Linking the Dispersion-‹Focalization Theory (DFT) and the Maximum Utilization of the Available Distinctive Features (MUAF) principle in a Perception-for-Action-Control Theory (PACT). M.J. Solé, P.S. Beddor, M. Ohala. Experimental Approaches to Phonology, Oxford University Press, pp.104-124.

Notes on the development of speaking styles over decades – the case of live football commentaries

J. Trouvain

Phonetics, Saarland University, Saarbrücken (Germany)
trouvain@coli.uni-saarland.de

Abstract: The aim of this paper to retrace the development of speaking styles within the genre 'live football commentary' by exploring several examples of broadcasts for a German audience in a time span of 80 years. It was shown that the affective information is consistently reflected by a pitch rise between one and two octaves during the phases of building up suspense before a goal and with the goal comment often at pitches beyond 400 Hz. Further genre-specific prosodic characteristics include pausing and articulation rate, which can be very different between television and radio broadcasters.

1. Introduction

1.1. Aims of the paper

Two main functions of prosody are used in a particular way in live football commentaries (henceforth LFCs): the expression of affective information (including attitude) and the construction of spoken genres (cf. Tench, 1996). For affective information prosody and the tone of voice during speech play very important roles. In addition, affect bursts (or raw interjections) can serve this function in an effective way. The prosodic shapes of various genres, styles and registers are bound to speech and have developed over time in their various forms. Regarding the genre LFC, the stylization of the 'goal roar' can be very different between cultures and languages: many Latin American commentators will use an extremely long vowel in the word "goal", often exceeding 10 seconds and longer, whereas such an extraordinary vowel lengthening would be considered as eccentric and overexcited and thus as completely inappropriate in other countries. It is the aim of this paper to retrace the development of speaking styles within the genre LFC by exploring several examples of broadcasts in a time span of 80 years.

1.2. "Invention" of live commentaries

The genre 'live football commentary' is a type of spoken text that came up with the distribution of the radio (starting in the 1920s) and later also in television (starting mainly in the 1950s). Before LFCs became the spoken genre with the most widespread audience for television (TV) in the world (reaching billions of people for a final of a men's world cup) the prosodic characteristics of LFCs changed over time. In this paper the aim is to show the development of LFC characteristics from before the 1950s up to the current decade.

The development of FLCs and other forms of radio broadcasting is closely linked to the invention, the distribution and the usage of the microphone. In the pre-radio days commentaries in general were not made for masses of people who were not witnesses of the commented events. The type of spoken interaction to other humans in live commentaries was something new with the microphone as clearly visible technical requirement for the transmission of the spoken information. Thus, speaking into and to a microphone required a completely new way of preparedness, in the case of *live* commentaries often linked with formulations where a preparation is restricted. In the beginning radio days the responsible persons developed some guidelines for "microphoning" (in German "für das Mikrophonieren", see Würzburger (1931) cited after Bose et al. 2013) where the well educated speakers, usually trained actors, got criteria how to speak in a microphone for radio broadcasting.

1.3. Stylistic diversity of microphone speech

Nowadays, in many cultures the use of microphones can be taken for granted for all people, e.g. in telephones. However, the great majority of microphone usage in a huge variety of technical devices concerns the interaction with one (or few) specific persons. These situations of microphone use can be considered as habitualised. This habitualisation of microphone use is in contrast to acoustic recordings for the purpose of scientific analysis in speech communication research. The situation are often unusual for the speakers, the majority of the recordings in the lab make use of prepared read speech and some recordings try to evoke naturalistic speaking conditions of spontaneous and conversational speech. Generally we are confronted with a great diversity of speaking styles where it is often unclear how much of the observations can be influenced by the recording situation (Wagner, Trouvain & Zimmerer 2015).

It can be seen as established that a recording device like a microphone has an influence on the way people speak and use language (when people are aware of the recording). The 'observer's paradox' postulated by Labov (1972) is probably reached at a high degree when untrained speakers with no experience in media communication are asked to speak in a radio or TV broadcast, particularly in a live situation where no cut-out of less perfect parts of the recorded speech is possible. With this paradox of microphone use in mind the professional speakers of LFCs have to do their speaking job for a very huge amount of people making a LFC an acted and an authentic style at the same time.

Live commentaries in sports can display a high level of affective load. The "live" character determined by unpredictable actions on the field and expressed with a "lively" voice of the usually male commentator. In contrast to TV commentaries those broadcasted by the radio have very clear audience design - the commentators can be considered as the cameramen of listeners, and they always give an interpretation of what they see and they consider reportable for the audience.

1.4. Further prosodic styles

The way soccer results are announced in Great Britain is an excellent example of a prosodic design of a speaking style. Bonnet (1980) showed that the information whether the first team won, lost or neither of both was clearly perceivable by the intonation for a huge majority of listeners. If the first team is the winner there is a rising contour on the score, if it is the loser it got a falling contour on the score, whereas a draw is signalled by a limited rise. Although this way of announcement was obviously introduced and popularised by one specific speaker who acted as a role model and 'invented' this style, raised pitch as a signal for victory is also used by commentators in goal jubilation in LFC (Trouvain 2011). Thus, the choice for the pitch contours are not completely arbitrary as already pointed out by Bonnet (1980).

Particular prosodic constructions are also used in horse race commentaries. An analysis of three races in Trouvain & Barry (2000) revealed that the English speaking commentators raised their pitch range in a stepwise fashion (similar to organ stops/registers) over the time course of the race. The effect for the listeners is that the suspense and drama continuously rise with the pitch of the voice. Contrary to the auditory impression of an increased speaking tempo the measured articulation rate was kept constant and the number of pauses increased (sic!). However, the pauses were much shorter towards the end and nearly every pause contained an audible breath sound. It can be assumed that audible breathing in short pauses evokes the impression of speaking fast and at a high degree of arousal.

Train station announcements belong to spoken genres people are confronted with in everyday life. Gilles (2014) could show with German data how prosody is used for this ritualised activity. Characteristic features are a very high pitch at the beginning of a prosodic phrase, a contrast between phrase-initial speeding up and phrase-final slowing down, low pitch accents, many de-accentuations, specific rhythmical patterns in lists, and segmental hyperarticulations. Sometimes de-accentuation typical for this style runs contrary to information to be conveyed. Particularly in such a decoupling of prosody from the content the ritualisation of this genre manifests itself, (see Gilles 2014: 13).

2. Styles in live football commentaries

2.1. Factors of variation

As Kern (2014) pointed out there are various phases of a game which can be distinguished also prosodically: *narration*, *pre-dramatic phases*, *supsense* and also the *climax* (goal presentation). Suspense and climax form together the dramatic phase. There are several further factors to be taken into account when looking at the prosodic variation of LFCs.

- Favourization of the team: a goal *for* the favoured team will be commented with more arousal than a goal *against* the favoured team.
- Importance of the game: a goal in a friendly game has less importance than in a final in the world cup and will be commented with less arousal.
- Importance of the goal: a 1-0 is more important and has more newness and reason for arousal than, say, a 7-0 with no suspense regarding the result of winning or losing.
- Medium: a radio commentator has to produce more words (and fewer and shorter pauses) to inform the listeners than a television (TV) commentator who just adds verbal information to the pictures.
- Culture: a goal comment in Latin America is different from the comment of the same goal of a German commentator.
- Situation in the game: a goal after an observable attack over seconds has a different suspense-building phase than an unexpected goal.
- Broadcast station: there are tendencies of speakers working for commercial stations to be prosodically more extreme than colleagues of public stations (Walhurst et al. 2013).
- Individuality: prosodic forming is one of the main tasks when performing on stage and during other forms of public presentations. This leads to prosodic differences between speakers.

2.2. Role models

Television broadcast was developed later than radio broadcast and also its popularisation took place at a later time. Consequently, the commentator for TV had as a role model the commentator for radio. However, as pointed out above, the TV commentator works differently than a radio commentator (Trouvain 2011). That means we have two strings of LFCs which have possibly developed independent from each other.

Similarly, a divergence between commentators of commercial and public broadcast stations could possibly have taken place. The commercial broadcasting speaker had as a role model the public broadcasting speaker (if public stations were first). In this case the divergence cannot be deduced to the different audience design of the TV speakers. It can be assumed that the "newer" speakers intend to stage differently, for instance by a more extreme prosody, in order to get more perceptual attraction.

3. Examples

The following examples serve as illustrations for some landmarks in the development of the prosodic styles in LFCs. The main focus is on the scenes that lead to a goal which includes the three phases of i) the pre-dramatic section, ii) the building up of suspense, and iii) the goal comment or goal roar (climax). The determination of the suspense phase was based on the beginning of a perceivable pitch rise which has been shown to be a general indicator (Samlowski et al., subm). All examples are from male speakers commenting for men's football games for a German speaking audience.

3.1. 1930

The example from 1930 is a radio commentary of German versus Italy (result 0-2) that took place in Frankfurt as a friendly game. The commentator uses over-long pauses (more than 2 secs). For the first goal the time between the pre-dramatic phase and the goal consists of nearly three seconds pause without building-up of any suspense. For the now suspense phase of the second goal the pitch rises from 160 to 350 Hz within a single phrase (13 syllables with a high articulation rate) which ends with an affect burst followed by many seconds of silence. The goal comments showed maximum pitch values of 398 and 439 Hz.

An interesting side observation from this 3 minutes excerpt of the entire commentary is that the speaker is not only addressing the unknown masses of public listeners but that he also integrate his comments in conversations with his neighbours in the stadium. These changes also lead to changes of the speaking register in terms of pitch register with a lowered pitch range for the non-distant addressees.

3.2. 1954

The examples from 1954 are two radio commentaries of the world cup final between Western Germany and Hungary (result 3-2). Note that the games between 1945 and 1989 were commented by the public broadcast stations of both German states in this period, the Federal Republic of Germany (FRG) and the German Democratic Republic (GDR).

The first example is taken from the Western German radio commentary on the decisive goal at this world cup final. This audio sample is well known in Germany because it is often used in German television as an index for a transitional point for an economic boost after the post-war years in (Western) Germany. However, the live broadcast of the TV did not take place with this audio comment. The recording from the radio was combined with the film that was broadcasted 1954. In the 1970s, when the new manipulated version was created, the audio information of the film was lost - in contrast to the audio of the radio broadcast. The more lively style of the radio comment superbly suits to underline the journalistic message of the "comeback" of (one half of) a national society.

The liveliness of the Western commentary can be nicely illustrated with the suspense phase which starts at 140 Hz and ends nearly 9 seconds later at 420 Hz, thus spanning a range of two octaves. The goal roar consisting of reduplications of "Tor" (German word for "goal") reaches a pitch of 446 Hz. The articulation rate for the pre-dramatic as well as for the suspense phase is in the range of normal conversations (4.6-4.9 syll/sec).

A very similar picture appears with the commentary from the Eastern German radio station. The suspense phase starts at 216 Hz, but takes with about 5 seconds less time than the colleague from the West. The Eastern German speaker shows even higher extreme pitch values: 458 Hz and the end of the suspense phase and 467 Hz for his production of "Tor".

Thus, both commentaries are at comparable levels, though the Eastern commentary shows more extreme values. It might be the case that this commentator considered the German team also as the own team.

3.3. 1974

During the world cup 1974 in Western Germany there was also a match between the Federal Republic of Germany and the German Democratic Republic (result 0-1). The examples are from TV in both states and from radio in Eastern Germany.

Pre-dramatically the commentator of the radio shows both, a phrase with rather many words followed by a rather long pause. The suspense phase is again characterised by many words (with 23 syllables) and only one pause but with a continuous rise from 178 to 286 Hz. The goal roar is at 321 Hz.

For the speaker of the Eastern TV the pre-dramatic comment is completely missing but the suspense phase is similar to horse race commentaries: there is a stepwise climbing of the pitch ranges from phrase to phrase from 170-233 Hz in phrase 1, to 216-330 Hz in phrase 2, and up to 315-382 Hz in phrase 3 where the extreme is reached with 382 Hz for the goal roar. The phrases are between 1 and 3 syllables long and the pauses are between 150 and 800 ms long.

The pre-dramatic phase of the Western TV commentator is similar to those found for radio commentators with many words, a high articulation rate and slightly longer pauses. The suspense phase with four phrases (between 1 and 4 syllables long) has a similar stepwise pitch structure as for the other TV commentary. But this speaker starts with a higher register: 231-289 Hz (phrase 1), 292-382 Hz (phrase 2), 331-405 Hz (phrase 3), 378-395 Hz (phrase 4). The goal comment (against the own team) is just "Tor" with 437 Hz as the highest value, followed by silence (but with the acoustic atmosphere in the stadium).

3.4. 1990

The 1990 examples are the radio and TV commentaries of (now reunified) German public broadcast stations. The commented goal is the decisive 1-0 in the world cup final between Germany and Argentina. It is important to note that this goal was shot by a penalty that was preceded with intensive discussions and no action on the field for several minutes. For the commentators it was not the usual way of building up suspense and they also had time to prepare the comments after the goal.

The goal roars were again at extremely high values (351 Hz in radio, 392 Hz in TV). The unusual situation of preparedness leads to longer narrative comments, also for the TV commentator. Interestingly, we can observe now a strong declination: the TV commentator decreases his pitch from 344 Hz to 173 Hz in a discourse unit of a bit more than 13 secs. The radio commentator's declination ranges from 320 to 163 Hz (in about 16 secs) before giving up his turn to his co-commentator.

3.5. 2010

One game in the round of 16 of the world cup 2010 was Germany v. England (result 4-1). Three example commentaries for the 1-0 (as a new and important goal) were selected: from public TV, public radio and from a commercial radio station.

All three commentaries are very similar regarding pitch during their goal roars and during the suspense phase (around 3 secs): TV from 123-393 Hz, climax at 439 Hz; public radio from 180-381 Hz, climax at 414 Hz; private radio from 160-450 Hz, climax at 426 Hz. The main difference between TV and radio commentators lies in the length of articulatory stretches. The TV commentator uses just 7 syllables for the suspense phase in contrast to 19-21 syllables by the radio commentators. The latter also use very high articulation rates between 6.8 and 7.3

syll/sec. Pausing is different, too: the TV commentator uses two very long pauses during the suspense, the public radio three short pauses whereas the private radio speaker do not pause at all.

4. Discussion

The suspense part seems to work in all inspected LFCs in the same way by a continuous or stepwise elevation of the pitch registers by between one and two octaves. The climax is usually but not in all cases reached during the goal comment. Although there is a large inter-individual variation between 321 Hz (East German radio 1974) and 467 Hz (East German radio 1954) it is astonishing that voices of male adults reach extremes usually known for baby cries or in most extreme situations of stress.

TV commentators share some characteristics such as using fewer words and more pauses in general. But they also apply longer pauses, particularly after the goal roar. Interestingly in two cases the suspense phase was completely omitted by the speaker, one case was from the very early radio days of LFCs.

What we can see from our examples is that the affective information displayed in the dramatic phases of 'building-up suspense' and the 'climax' is very similar in these styles over 80 years. The possibilities of the expression of arousal by pitch seem to be limited (cp. the suspense phase of horse race commentaries described in Trouvain & Barry 2000), at least in the inspected German data. More possibilities for changes in prosodic styles for this genre appear to be in pausing and articulation rate. Both temporal parameters were quite variable across the samples.

It is hard to draw any further conclusions from the rather small database of 12 commentaries investigated here. Thus, an extension of the data collection would be the next step. Such an extension could also include new forms of LFC such as audio descriptions for the blind (Trede 2007) where the commentator is rather close to his audience in the stadium and delivers slightly different information than in a radio broadcast for the time of the entire game.

Acknowledgement

The author would like to thank Frank Zimmerer for helpful comments on an earlier draft of this paper.

References

Bonnet, G. 1980. A study of intonation in soccer results – Wolverhampton Wanderers 2 Nottingham Forest ? Journal of Phonetics 8, 21-38.

Bose, I., Hirschfeld, U., Neuber, B. & Stock, E. 2013. Einführung in die Sprechwissenschaft. Tübingen: Narr.

Gilles, P. 2014. Zur Prosodie von Bahnhofsansagen. In: K. Birkner, P. Bergmann, P. Gilles, H. Spiekermann, & T. Streck (eds.). Sprache im Gebrauch: räumlich, zeitlich, inter-aktional: Festschrift für Peter Auer. Heidelberg: Universitätsverlag Winter, 95-108.

Kern, F. 2010. Speaking dramatically: The prosody in radio live commentaries of football games. In: Selting, M., Barth-Weingarten, D. & Reber, E. (eds): Prosody in Interaction. Amsterdam: Benjamins, 217-238.

Labov, W. 1972. Some principles of linguistic methodology. Language in Society, 1, 97–120.

Samlowski, B., Kern, F. & Trouvain, J. submitted. Perception of suspense in live football commentaries from German and British perspectives.

Tench, P. 1996. The Intonation Systems of English. London: Cassell.

Trede, B.-J. 2007. Ich sehe was, was Du nicht siehst. Fußball-Live-Reportage für Blinde und Sehbehinderte: Inhalte, Funktionen und Perspektiven einer jungen journalistischen Darstellungsform. In: Settekorn, W. (ed) Fußball – Medien, Medien – Fußball. Zur Medienkultur eines weltweit populären Sports. Hamburger Hefte zur Medienkultur, Universität Hamburg.

Trouvain J. 2011. Between excitement and triumph – live football commentaries in radio vs. TV. Proc. of the 17th International Congress of Phonetic Sciences, Hong Kong, 2022-2025.

Trouvain, J. & Barry, W.J. 2000. The prosody of excitement in horse race commentaries. Proc. of the ISCA-Workshop on "Speech and Emotion", Newcastle (N. Ireland), 86-91.

Wagner, P., Trouvain, J. & Zimmerer, F. 2015. In defense of stylistic diversity in speech research. Journal of Phonetics 48, 1-12.

Walhurst, S., McCabe, P., Yiu, E., Heard, R. & Madill, C. 2013. Acoustic characteristics of male commercial and public radio broadcast voices. Journal of Voice 27(5), 655.e1-655.e7.

Würzburger, K. 1931. Die Erziehung zum Redner. Die Sendung 34, 670-671.

Author Index

www.ingramcontent.com/pod-product-compliance
Lightning Source LLC
Chambersburg PA
CBHW081535220326
41598CB00036B/6446

Studientexte zur Sprachkommunikation
Herausgegeben von Rüdiger Hoffmann
ISSN 0940-6832

Band 79:
Rüdiger Hoffmann, Jürgen Trouvain (Eds.): HSCR 2015

ISBN 978-3-95908-020-0

The international workshop on the "History Speech Communication Research" is organised by the Special Interest Group (SIG) on "The History of Speech Communication Sciences". This SIG is supported by the International Phonetic Association (IPA) and the International Speech Communication Association (ISCA). The workshop offers an exchange forum for researchers with work on all kind of historical aspects of the research fields represented at the Interspeech conferences and the Congresses of Phonetic Sciences (ICPhS).

This volume includes the proceedings of the first workshop of this series, held in Dresden as a satellite event of Interspeech 2015. The re-opening of the historical acoustic-phonetic collection (HAPS) in new rooms was part of the workshop. With the Barkhausen building on the first day of the workshop and the technical collections of the city of Dresden in the historical Ernemann building on day two of the workshop, the meeting had appropriate locations.

The proceedings volume contains 18 contributions, including the general presentations of the opening session as well as the papers from the dedicated sessions on "mechanical speech synthesis", "collections", and "pioneering work in phonetics". The papers are written from 23 authors, coming from nine countries.

TUD_press_

Wilhelm Caspary | Klaus Wichmann

Auswertung von Messdaten

Statistische Methoden für Geo- und
Ingenieurwissenschaften

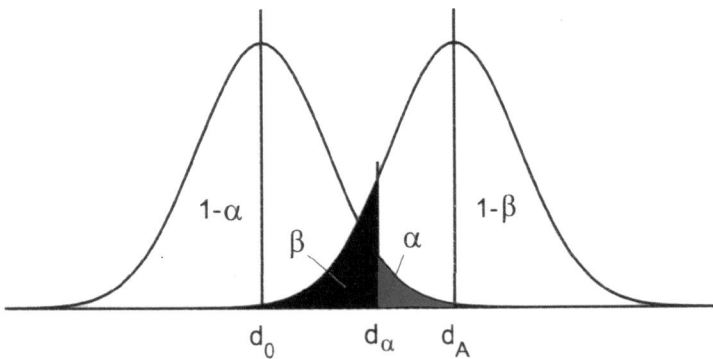

Oldenbourg